UNIFIED S-BAND TELECOMMUNICATIONS TECHNIQUES FOR APOLLO

VOLUME I

FUNCTIONAL DESCRIPTION

By John H. Painter and George Hondros

Manned Spacecraft Center
Houston, Texas

NATIONAL AERONAUTICS AND SPACE ADMINISTRATION

FOREWORD

The Review Copy of this publication was entitled, "The Apollo Unified S-Band Telecommunications System - Volume 1", and identified as File Number S-58 by NASA Manned Spacecraft Center. The present publication, under a new title, is nearly identical to the Review Copy with only a few minor changes.

The authors wish to acknowledge the aid and cooperation received in the preparation and review of this document.

    W. Victor
    M. Easterling
    B. Martin   Jet Propulsion Laboratory
    L. Randolph   Pasadena, California
    P. Goodwin
    M. Brockman

    D. Holcomb   Motorola, Inc.
                             Scottsdale, Arizona

TABLE OF CONTENTS

Section		Page
1.0	INTRODUCTION	2
	1.1 Objectives of the Report	2
	1.1.1 Volume I	2
	1.1.2 Volume II	2
	1.1.3 Volume III	3
	1.2 History of the System	3
2.0	THE BASIC SYSTEM	6
	2.1 System Concept	6
	2.1.1 The Basic Spacecraft System	6
	2.1.2 The Basic Ground System	7
	2.2 Extension of the Basic System Required to Support the Command Service Module and Lunar Excursion Module	8
	2.3 The Ground Network	8
3.0	DETAILED SYSTEM DESCRIPTION	10
	3.1 The Communications Signals	10
	3.1.1 Up-link (Ground to Spacecraft)	10
	3.1.2 Down-link (Spacecraft to Ground)	10
	3.2 Signal Transmission Modulation Techniques	12
	3.2.1 Up-link	12
	3.2.2 Down-link	12
	3.2.3 Signal Design	14
	3.3 Spacecraft Subsystem Configuration	16
	3.3.1 Premodulation Processor	16
	3.3.2 Transponder	19
	3.4 Ground Subsystems	22
	3.4.1 General Configuration	22
	3.4.2 Antennas	23
	3.4.3 Microwave Circuitry	23
	3.4.4 Reference Channel Receiver	24

Section			Page
	3.4.5	Angle Channel Receiver	25
	3.4.6	Transmitter	25
	3.4.7	Ranging Circuitry	26
	3.4.8	Demodulation Circuitry	28
	3.4.9	Premodulation Circuitry	31
	3.4.10	Acquisition and Programing Circuitry	31
	3.4.11	Ground System Peripheral Equipment	31
4.0	SYSTEM OPERATIONAL TECHNIQUES		34
	4.1 Ranging and Spacecraft Acquisition		34
		4.1.1 A Physical Explanation of the Ranging Process	34
		4.1.2 An Explanation of Acquisition	36
	4.2 Lunar Mission Communications Requirements		37
		4.2.1 Pre-Launch	37
		4.2.2 Launch	38
		4.2.3 Earth Orbit	38
		4.2.4 Translunar Injection	38
		4.2.5 Spacecraft Transposition and LEM Checkout	38
		4.2.6 Earth-Lunar Coast	38
		4.2.7 Lunar Orbit	38
		4.2.8 Lunar Landing	39
		4.2.9 Moon-Earth Injection and Coast	39
		4.2.10 Reentry	39
	4.3 System Operation for a Nominal Lunar Mission		39
		4.3.1 Pre-launch	39
		4.3.2 Launch	39
		4.3.3 Earth-Orbit Insertion	40
		4.3.4 Earth-Orbit	40
		4.3.5 Trans-Lunar Trajectory Injection	40
		4.3.6 Transposition and LEM Check-out	40
		4.3.7 Translunar Orbit	41
		4.3.8 Lunar Orbit	41
		4.3.9 Lunar Landing, Surface Operations and Rendezvous	41
		4.3.10 Earth Trajectory Injection and Coast	42
		4.3.11 Atmospheric Reentry	42
5.0	REFERENCES		43
6.0	TABLES		44
7.0	FIGURES		

UNIFIED S-BAND TELECOMMUNICATIONS TECHNIQUES FOR APOLLO

VOLUME I

FUNCTIONAL DESCRIPTION

By John H. Painter and George Hondros
Manned Spacecraft Center

SUMMARY

This document is a functional description of the intended configuration and operation of the Apollo Unified S-band Telecommunications and Tracking System. In particular, this description applies to the link between the command-service module and the ground tracking station. The lunar excursion module-to-ground link is similar, and an equivalent description applies.

The document begins with a short resume of the system's developmental history. Then the basic spacecraft and ground system concept are simply explained. Following are detailed descriptions of the communications signals, modulation techniques, and subsystem configurations. The document concludes with explanations of system operational techniques.

1.0 INTRODUCTION

1.1 Objectives of the Report

The Unified S-Band Telecommunications and Tracking System under development for the Apollo lunar missions is not now well documented. Although some documentation on the individual subsystems exists, there is no complete system documentation which provides a reader with the basic system design, capabilities, operation, and limitations. This report is intended to provide such system documentation for the system in its current state of development and is divided into three volumes because of the system complexity.

Detailed description of other elements that may feed into the present system but are not directly developed by MSC is planned to appear in later volumes of this report or in appropriate reference materials as they are developed.

1.1.1 Volume I

Volume I provides a simplified functional description of the configuration and operation of the communication and tracking system to be used in the lunar Apollo missions. The spacecraft S-band subsystems for the command-service module (CSM) and lunar excursion module (LEM) are very similar in a system sense. Both vehicles will, in fact, be supported by the same type of ground station communication equipment. Therefore, the basic spacecraft description, given here, is for the CSM, with the minor differences of the LEM subsystem pointed out. The basic system modulation techniques and operational concepts are described.

At the time Volume I was prepared, the CSM spacecraft S-band subsystem was being redesigned to accommodate additional communication requirements, insure equipment reliability, and insure compatibility with the ground subsystem. The spacecraft subsystem configuration chosen to be described in this volume is one which resulted from detailed technical discussions between Manned Spacecraft Center, Jet Propulsion Laboratory, NASA Headquarters, Goddard Space Flight Center, and associated contractors. Although the authors feel that this configuration best meets the presently defined communication requirements and provides more flexibility than other presently proposed configurations, it is possible that the final spacecraft S-band equipment may differ slightly (at the module level) from the one described. It is not expected that the basic design philosophy or operation of the spacecraft S-band subsystem will depart appreciably from that described.

At the time of preparation of Volume I, the ground subsystem demodulators were functionally specified, but not designed. The authors have treated the set of demodulators which, in their judgment, best satisfy the requirements.

1.1.2 Volume II

Volume II is a detailed mathematical analysis of the communication and tracking channels of the system. Equations in generalized parameters are derived which may be used either to design the communication channels or to analyze existing designs. The equations are sufficiently general that changes of the type referred to in section 1.1.1 may be easily treated. Basic assumptions made in the derivations are plainly stated so that the validity of the channel models may be easily inferred.

1.1.3 Volume III

Volume III is a tabulation of performance data or circuit margins for each channel of the system. The calculation of these circuit margins is based on the theoretical analysis presented in Volume II. The circuit margins in Volume III allow inference of the quality of the various system channels during nominal as well as non-nominal modes of transmission of the spacecraft and ground systems.

1.2 History of the System

The present configuration of the Unified S-Band has evolved from earlier equipment. It is therefore important to acquaint the reader with the history of the system. The Mercury spacecraft was provided with many electro-magnetic transmitting and receiving systems. These systems operated at seven discrete frequencies within five widely separated frequency bands. The systems used on the Mercury spacecraft were:

1. HF voice transmitter and receiver

2. UHF voice transmitter and receiver

3. VHF telemetry transmitter no. 1

4. VHF telemetry transmitter no. 2

5. C-band transponder

6. UHF command receiver

7. S-band transponder (pulse)

To complement the spacecraft systems, a number of ground stations, strategically located about the globe, were provided with systems compatible to those in the spacecraft to fulfill the communications and tracking requirements for the Mercury missions. The spacecraft and ground systems used in the Mercury Project performed satisfactorily. As a result, it was planned that the great majority of these systems should be used in the Gemini Project, with, perhaps, some systems modifications made to provide better performance or power and weight savings in the Gemini spacecraft.

When the Apollo Project was initiated, it was stipulated that as much as possible of the existing Mercury ground network and spacecraft systems be used. In addition to these systems, it was conceived that a transponder should be included in the spacecraft to perform the ranging operation at lunar distances. The transponder was also to be used for transmission of voice and telemetry at lunar distances. Since the transponder design chosen was compatible with the

Deep Space Instrumentation Facility (DSIF), established by the Jet Propulsion Laboratory (JPL), the JPL technique (pseudo-random code ranging) was chosen by NASA to perform ranging. Thus, for the deep space phase of communications for the lunar mission, the JPL transponder was to perform the communication and tracking functions using three deep space stations, closely resembling the JPL design. For the near-earth phase, however, the VHF, UHF, and C-band Gemini-type equipments were to be used to perform the communications and tracking functions. A study[1] conducted by JPL, however, indicated that "under worst conditions" the deep space stations might not acquire the spacecraft at altitudes lower than 10,000 nautical miles[2]. In addition, computations made by the contractor and MSC indicated that the VHF and UHF systems range capability would be less than 10,000 nautical miles.

During the early phases of spacecraft subsystems design, performed by the contractor, it was realized that a problem with the spacecraft weight would arise, since the near-earth phases of the mission required a number of different spacecraft transmitters and receivers, as well as their back-ups. Because of this realization, the contractor suggested to NASA that additional consideration should be given to the spacecraft weight problem, and that perhaps some other communications and tracking system could be used during the various near-earth phases of the Apollo lunar missions.

As a result of the foregoing considerations, a meeting was held at the Office of Tracking and Data Acquisition Systems (OTDA) in Washington, D.C. in December 1962. At this meeting OTDA presented plans for a ground network using a unified S-band carrier system to representatives of various NASA centers, including JPL and MSC. The ground portion of the system was to consist of three stations having 85-foot Cassegrain feed antennas for deep space communications and tracking (separate from those of JPL), and a number of stations with 30-foot Cassegrain feed antennas to perform communications and tracking during the near-earth phases of the Apollo lunar missions. This proposed ground network not only would increase the range capabilities for near-earth communications and tracking, but would also allow transmission from the spacecraft during both near-earth and deep-space phases to be performed by one transmitter, thus eliminating all the VHF, C-band and UHF systems and their back-ups, and consequently reducing the spacecraft weight.

The philosophy of using the unified S-band system for communications and tracking at near-earth and in deep space has been accepted. As a result, the ground network required for the support of lunar missions is presently being implemented under a contract from the Goddard Space Flight Center.

---

[1] EPD-29 - Estimated 1963-1970, Capability of the Deep Space Instrumentation Facility for Apollo Project, JPL, Pasadena, California, 2-1-62.

[2] The altitude of 10,000 nautical miles was calculated on the basis of the three JPL DSIF stations located at Goldstone, California, Johannesburg, Union of South Africa, and Woomera, Australia.

The Apollo unified S-band system has not yet been evaluated in the laboratory, since the hardware design and fabrication has not been completed. Plans have been prepared, however, by the Manned Spacecraft Center for early and continuing laboratory testing as well as flight qualification of the system.

2.0 THE BASIC SYSTEM

2.1 System Concept

The primary concept underlying the Apollo S-band telecommunications system was originally that all communications and data transfer between spacecraft and ground should be made using one common set of equipment and one radio frequency carrier for transmission. Because of various historical and mechanical reasons, this concept has not been fully implemented in the Apollo system. The historical reasons have been mentioned in the previous section. Because of the ground requirement that the spacecraft transmit a stable carrier spectral line, phase-coherent with the up-link carrier, to enable two-way Doppler tracking, ranging, and ground antenna pointing, narrow deviation phase modulation has been implemented for most of the spacecraft transmission. Those information functions requiring a large modulated bandwidth are modulated directly on the carrier while other, less wide, functions are first modulated onto subcarriers. Since the range code and television signals transmitted by the Apollo spacecraft both require large modulated bandwidths, and, in some instances, will be transmitted simultaneously, it was necessary to place them on separate carriers, using phase modulation for the first and frequency modulation for the second. Aside from this exception, however, the basic concept of one carrier and one set of equipment has been closely approached in the Apollo system. Failure and redundancy considerations have, of course, required duplicate standby equipments, both in the spacecraft and on the ground. The system described in this report will be used in the translunar and lunar phases of the Apollo mission and during the injection phase from earth orbit to lunar transfer trajectory. It is planned that the system will be used in the earth orbital mission phases also.

The heart of the unified Apollo system is the pseudo-random code ranging subsystem developed by Jet Propulsion Laboratory. This system uses a ground-based transmitting and receiving station working in conjunction with a spacecraft transponder. A pseudo-random code is phase modulated on an S-band carrier at the ground and transmitted to the spacecraft. The code modulation is recovered in the transponder and retransmitted to the ground station on a different S-band carrier which is phase coherently generated from the up carrier. At the ground station the time difference between the transmitted and received code gives a measure of the spacecraft range.

2.1.1 The Basic Spacecraft System

Neglecting redundancy, the basic spacecraft system is shown, highly simplified, in figure 1. The system is composed of a premodulation processor, transponder, final amplifier and associated microwave circuitry, high-gain antenna and omnidirectional antenna. This system diagram is later expanded (section 3.3) to show the means for obtaining switchable redundancy and simultaneous PM-FM operation.

The premodulation processor accepts voice, PCM telemetry, biomedical data, television, and emergency key signals from the spacecraft for transmission to the ground. It also recovers the voice and up-data signals received from the ground.

The CSM transponder is basically a narrow band PM receiver, narrow band PM transmitter exciter, and wide band FM transmitter exciter. The LEM transponder, however, does not contain an FM exciter since it is not required for the LEM to transmit FM and PM information, simultaneously. The transponder feeds and is fed by a package containing final amplifiers and microwave switching and diplexing circuitry. The microwave circuitry feeds the two spacecraft antennas which are manually selectable. The PM receiver has a local oscillator, phase locked to the received carrier, which provides the frequency and phase reference for the PM transmitter exciter. The FM transmitter exciter has its own separate oscillator. The PM receiver is used at all times to receive from the ground. The PM exciter is used at all times to transmit to the ground. The FM exciter is used only when transmission of FM data is required.

2.1.2 The Basic Ground System

Again neglecting redundancy, the basic ground system is shown, highly simplified, in figure 2. The system is composed of a high-gain main antenna, wide-beam acquisition antenna, microwave circuitry, a main reference channel receiver, acquisition reference channel receiver, two main angle channel receivers, two acquisition angle channel receivers, a transmitter, data demodulation circuitry, ranging circuitry, premodulation circuitry, acquisition and programing circuitry, data handling equipment, and peripheral equipment.

The acquisition channels, transmitter, and acquisition antenna are used initially to acquire the spacecraft signal. This operation consists of a search in angle with the acquisition antenna and a search in frequency with the acquisition reference channel receiver for the central PM carrier component of the spacecraft signal. The ground receiver local oscillator phase locks to the received carrier, thus activating the angle channels. When the acquisition antenna is sufficiently well aligned, the main antenna, which is physically tied to the acquisition antenna, acquires the spacecraft carrier. The main reference channel receiver is then phase-locked and the main angle channels become effective. The drive for the antenna servos is then switched from the acquisition to the main angle channels.

The ranging circuitry contains digital equipment for generating the various range codes and range measurements, Doppler measuring circuitry, and a range code receiver which is fed by the reference channel 10-megacycle IF outputs. The ranging circuitry feeds the range code to the transmitter phase modulator, where it is effectively summed with other up-going data from the premodulation circuitry.

The data demodulator, as shown in figure 2, accepts PM data from the main reference channel receiver and FM data from the acquisition reference channel. Since the two reference channel receivers are identical, the acquisition reference channel is available, after completion of acquisition, for the reception of other data, namely FM. The data demodulation and ranging equipment both feed the data handling equipment. The data handling equipment also feeds the premodulation equipment.

The acquisition and programing circuitry plays an essential role with the rest of the equipment. The ranging processes are automated, and are digitally programed in response to analog measurements on various parts of the equipment. The programing circuitry contains special purpose digital computing equipment and analog-to-digital converters. Additional peripheral equipment, not shown in figure 2, forms another part of the basic ground system. This equipment is treated in section 3.4.

2.2 Extension of the Basic System Required to Support the Command/Service Module and Lunar Excursion Module

The Apollo lunar mission uses two spacecraft, the command/service module (CSM) and lunar excursion module (LEM). The CSM communications system is active throughout the mission, from prelaunch through reentry and landing. The LEM communications system is active mainly in the region of the moon. The mission communication requirements for the two spacecraft are treated in detail in section 4.2.

The fact that there are two spacecraft, each having its own S-band system, complicates the ground station. Figure 3 shows a configuration for a ground station having a full duplicate ability to range simultaneously and communicate simultaneously with both spacecraft. This configuration implies that both spacecraft remain within the beamwidth of the high-gain ground antenna and that antenna pointing is performed on only one of the spacecraft. It is seen that to the basic ground station have been added two reference channel receivers, one data demodulator, one set of ranging circuitry, one transmitter and one set of premodulation circuitry. It should be understood that the capability of the data handling and acquisition and programing circuitry is also expanded over the basic system. Figure 3 still ignores redundancy, although the ability to switch data demodulators and ranging circuitry from one reference channel to another implies redundancy of a sort.

Functionally, the two spacecraft S-band systems are alike, although there may be differences in physical packaging and available output power levels. These differences are discussed further in the detailed system description, section 3.3.

2.3 The Ground Network

There are three types of S-band ground stations; the deep-space stations, injection and transposition gap-filler stations, and earth-orbital stations. The types of stations differ in the sizes of their antennas and whether they have single or dual tracking capability.

The three deep-space stations have 85-foot diameter antennas and full dual tracking and communication capability. These stations will be located at Goldstone, California; Madrid, Spain; and Canberra, Australia. Following the injection into lunar transfer trajectory and during the LEM transposition phases of the Apollo lunar mission, it is possible that the spacecraft will not have been acquired by a deep-space station. This possibility exists because the spacecraft, at the beginning of injection, is at a relatively low altitude

(100 nautical miles). Depending on the geographical relation of the injection point to the nearest deep-space station, the spacecraft may have to climb several thousand miles in altitude before becoming visible.

Because of this possible gap in coverage, several injection gap-filler stations, number and location now unknown, will be used to supplement the deep-space stations for those missions where deep-space station acquisition is delayed. These stations will have 30-foot antennas and dual data communication capabilities. The S-band ground stations to be used in the earth-orbital mission phases will have 30-foot antennas and single ranging and communications capabilities. The locations of the stations will be such as to provide good coverage of Apollo earth orbits, in conjunction with the gap-filler stations.

3.0 DETAILED SYSTEM DESCRIPTION

3.1 The Communications Signals

The following section describes the various communication, data, and ranging signals which serve as inputs to the system, either on the ground or in the spacecraft. The signals are described as they exist in their unmodulated or baseband form. The modulated signals are treated in section 3.2.

3.1.1 Up-link (Ground to Spacecraft)

3.1.1.1 <u>Range code</u>.- One of the most important signals generated by the ground station is the range code, used to determine the radial range between ground station and spacecraft. This code is a continuous running binary waveform which progresses at a nominal rate of a million bits per second, being generated from a 496-kilocycle clock signal. Being one of a class of codes known as pseudo-random, this code has a basic period of about 5.4 seconds. This period insures that the code does not repeat for 5.4 million bit periods. A period of this length is required so that the code will not recycle during the time of propagation from earth to moon and back.

The transmitted code, which is generated by digital logic circuitry, is a Boolean combination of four shorter codes and the digital clock signal (a 496-kc square wave). For general information the combining function is:

$$\text{Code} = \left\{x \cdot c\ 1\right\} \ v \ \left\{\overline{x} \cdot \left[(a \cdot b\ v\ b \cdot c\ v\ a \cdot c) \oplus c\ 1\right]\right\}$$

Where the dot indicates the Boolean "and", the v indicates the Boolean "or", the $\oplus$ indicates the Boolean "exclusive or", and the bar indicates the Boolean "not". The lengths of the code components in bit periods are given in table I. For detailed information on the generation of this code, see reference 1.

3.1.1.2 <u>Up-Data</u>.- It is required that a digital up-data link be available to transmit to the spacecraft (both CSM and LEM) either up-dating information for the on-board computer or real-time or stored-program commands. The signal brought in to the S-band equipment is derived from a Gemini-type Digital Command System (DCS). This signal is a composite waveform consisting of a 1-kilocycle sinusoid summed with a 2-kilocycle sinusoid which has been phase-shift keyed by a 1-kilobit digital signal. The digital signal is a basic 200-bit information signal, sub-bit encoded, 5 bits for 1.

3.1.1.3 <u>Voice</u>.- Voice transmission is also required to the spacecraft. The baseband voice signal is simply an analog waveform with most of its energy lying between 300 and 2,300 cycles per second. This signal originates from the various channels in the ground station.

3.1.2 Down-link (Spacecraft to Ground)

3.1.2.1 <u>Range code</u>.- The range code signal, recovered in the spacecraft for retransmission to the ground, is approximately the same as that generated on the ground, except that it is contaminated with noise and two up-link subcarriers (section 3.2). The code is running at a slightly different rate due to Doppler effects.

3.1.2.2 <u>Pulse-code-modulated telemetry</u>.- The primary spacecraft telemetry signal is binary PCM (non-return to zero) having either a 51.2-kilobit or 1.6-kilobit information transmission rate. The basic onboard generated telemetry clock frequency is 512 kilocycles.

3.1.2.3 <u>Voice</u>.- The baseband voice signal is much like the ground station voice signal. Clipping and filtering is employed to keep the average voice level high and most of the power between 300 and 2,300 cycles. The voice signal originates from the astronaut's microphone or the AM relay receiver used during extravehicular astronaut activities.

3.1.2.4 <u>Biomedical telemetry</u>.- Biomedical data from the astronauts is routed through the PCM telemetry when the astronauts are in the spacecraft. When an astronaut is outside the spacecraft, either in space or on the surface of the moon, a special set of biomedical telemetry channels is provided. Seven low frequency subcarriers in the astronaut's suit communications package are frequency modulated with the biomedical measurements. The modulated subcarriers are summed and amplitude modulated on a VHF carrier for relay back to the parent spacecraft. At the spacecraft the AM signal is demodulated and the summed frequency-modulated subcarriers are recovered. This recovered signal forms the baseband input to the S-band system. The center frequencies and channel information of the various biomedical subcarrier channels are given in table II.

3.1.2.5 <u>Television</u>.- During certain mission phases the spacecraft will transmit television to the ground. Because of spacecraft power limitations the television picture will not be of the usual broadcast quality. The television signal is analog, having a basic picture format of 10 frames per second, 320 lines per frame, with an aspect ratio of 4:3. The resolution is further limited by 500-kilocycle baseband low pass filtering. Amplitude synchronization will be used with the synchronizing level 30 percent above the black level.

3.1.2.6 <u>Emergency voice</u>.- A capability is required for successful voice communication with the ground in the event of failure of the spacecraft high-gain antenna or final amplifier, but not both. The emergency voice baseband signal is the normal voice, routed through a different channel.

3.1.2.7 <u>Emergency key</u>.- For a last-resort communication capability the astronauts may use hand-keyed Morse code. The baseband key signal is an on-off d-c voltage obtained by keying the spacecraft 28-volt battery.

3.1.2.8 <u>Recorded PCM telemetry</u>.- PCM telemetry at either a 1.6 kbs rate or 51.2 kbs rate may be recorded when necessary. When the recorded telemetry is played back into the transmission channel the playback signal has a nominal 51.2 kbs rate with, perhaps, some "wow" and "flutter" due to the tape recorder. The playback signal is routed through a different channel than the real-time telemetry.

3.1.2.9 <u>Recorded voice</u>.- The normal clipped voice may be recorded at any time. The recorded voice is played back at a higher rate than the recording rate (greater than 4:1). The playback signal occupys a greater base bandwidth than the normal voice signal (greater than 9 kc) due to the high playback rate. The playback signal is routed through a different channel than the real-time voice.

3.2 Signal Transmission Modulation Techniques

3.2.1 Up-link

The up-link modulation techniques are simpler, conceptually, than the down-link and so are treated first. The final modulation process on the outgoing carrier is phase modulation using relatively narrow deviation. Narrow phase modulation is required to insure that a phase stable carrier component arrives at the spacecraft. Since the spacecraft transmission carrier is derived phase coherently from the received carrier, it is important that the carrier received at the spacecraft not be allowed to reverse phase for any appreciable length of time due to modulation, since this reversal would cause an error in two-way Doppler tracking between the ground transmitted and received carriers. Narrow phase deviation also assures that the modulation is concentrated in the first order side products of the modulated signal. This concentration optimizes the particular reception technique employed in the spacecraft.

The total rms phase deviation on the up-link carrier is kept at about one radian. For a classical treatment of signal design, see reference 2.[1]

3.2.1.1 <u>Range code</u>.- The range code, which is a binary analog waveform, is effectively summed with the other up-going information and phase modulated onto the carrier.

3.2.1.2 <u>Up-data</u>.- The baseband up-data signal is first frequency modulated on a 70-kc subcarrier, then summed with the other up-going information and phase modulated on the carrier.

3.2.1.3 <u>Voice</u>.- The baseband voice signal is first frequency modulated on a 30-kc subcarrier, then summed with the other up-going information and phase modulated on the carrier.

3.2.2 Down-link

The down-link modulation techniques for CSM and LEM are practically identical. The CSM may transmit on two S-band carriers simultaneously. The LEM, however, may time share one carrier for transmission of FM or PM modulation. Considering both spacecraft, three separate carrier frequencies may be received at the ground, simultaneously. The various modulation parameters will be treated in the section on signal design.

NOTE: 1. The only difference between the up-links for CSM and LEM is the carrier frequency. The modulation techniques are identical. The various modulation parameters are treated in section 3.2.3, Signal Design.

There are two final modulation processes in the spacecraft. These processes are phase modulation and frequency modulation. Similar to the up-link, a phase-stable carrier component must arrive at the ground station for use in Doppler tracking and for use by the narrow angle channels which point the ground antenna. This carrier requirement implies narrow phase modulation for some of the information links. In addition, the carrier requirement implies that the LEM can not transmit FM modulation when tracking and ranging is required by the ground. As explained in section 2.1, a separate frequency-modulated carrier was chosen for the transmission of television and certain other data. Certain of the information signals can be transmitted only in the PM channel, and certain other signals can be transmitted only in the FM channel. Some of the information signals can be transmitted in either channel. The combination modes are spelled out explicitly in the signal design section 3.2.3.

3.2.2.1 <u>Range code</u>.- The range code, having been recovered in the spacecraft, is effectively summed with the other PM channel information and phase modulated onto the down carrier. The range code is restricted to the PM channel only.

3.2.2.2 <u>Pulse code modulated telemetry</u>.- The PCM telemetry signal, binary coded, is first phase modulated on a 1.024-mc subcarrier. The modulated subcarrier may then be either summed with PM channel information and phase modulated on the down PM carrier or summed with FM channel information and frequency modulated on the down FM carrier.

3.2.2.3 <u>Voice</u>.- The baseband voice signal is first frequency modulated on a 1.25-mc subcarrier. As with the PCM telemetry subcarrier, the voice subcarrier may be summed with other information and be transmitted through either the PM or FM channel.

3.2.2.4 <u>Biomedical telemetry</u>.- When employed, the biomedical telemetry baseband signal will first be summed with the baseband voice signal. Then both signals will be frequency modulated on a 1.25-mc subcarrier (the voice subcarrier). This modulated subcarrier may then be transmitted through either the PM or FM channel.

3.2.2.5 <u>Television</u>.- The baseband television signal is summed with other FM channel information and frequency modulated on the down carrier. Television is restricted to the FM channel only.

3.2.2.6 <u>Emergency voice</u>.- The baseband voice signal is phase modulated directly on the carrier using a sufficiently narrow deviation to insure the retention of a phase-stable carrier-spectral component for ground Doppler track.

3.2.2.7 <u>Emergency key</u>.- The keyed d-c signal is applied to a digital "and" circuit, to which is also applied the 512-kc square wave from the spacecraft central timing equipment. The output of the "and" circuit, a keyed 512-kc square wave, is then band-pass filtered to give a keyed 512-kc subcarrier. The keyed subcarrier is phase modulated on the carrier using a narrow enough phase deviation to maximize the first order side product and to insure retention of a phase-stable carrier-spectral component for ground Doppler track.

3.2.2.8 _Recorded PCM telemetry_.- The recorded PCM telemetry signal is bi-phase modulated on a 1.024 mc subcarrier. This is not the same subcarrier used for real-time telemetry. The modulated subcarrier may then be summed with other information and transmitted through either the FM or PM channel. The subcarrier for recorded telemetry cannot be transmitted through the same carrier channel as that carrying the real-time telemetry subcarrier.

3.2.2.9 _Recorded voice_.- The recorded voice signal is frequency modulated on a 1.25 mc subcarrier. This is not the same subcarrier used for real-time voice. Because of base-bandwidth considerations, the relayed EVA biomedical data cannot be summed with recorded voice. The modulated subcarrier may be summed with other information and transmitted through either the PM or FM channel. As with recorded telemetry, recorded voice and real-time voice cannot be transmitted through the same carrier channel.

3.2.3 Signal Design

It has been stated previously (sec. 3.2.2) that for the down link certain information functions are restricted to PM only or FM only, while some functions can be transmitted either way. There are many possible combinations of transmissions. The choice of the various modulation parameters for these combinations is referred to, here, as signal design. At this time, first cut signal design and optimization for both the up and the down links have been performed by the prime contractors for CSM and LEM, respectively.

3.2.3.1 _Combination modes_.- It is necessary to designate all the possible transmission modes, both for the up- and the down-links. This is done in table III for the up-link. Table IV does not list all the possible combinations for the down-link, but lists most of those of interest. Table IV, also indicates which modes use the spacecraft high power final amplifiers (PA) and which modes result for use of the PM exciter only.

3.2.3.2 _Modulation parameters_.- For the purposes of Volume I, the modulation parameters to be treated are subcarrier frequencies and various modulation indices. There are several separate sets of modulation indices for this system; those of the information on the subcarriers, and those of the subcarriers on the carriers. At this time contractors have derived sets of modulation indices. It is intended that Volume III will present a third set of indices, derived and optimized by NASA-MSC. Table V presents only the subcarrier frequencies, leaving the indices for Volume III.

3.2.3.3 _Modulated spectra_.- The modulation modes described in section 3.2.3 are best illustrated through plots of energy density against frequency. The spectra of interest are those for the individual up-links, individual down-links, and the dual up-links and down-links where all channels are active. The dual link plots show how the spectra for both spacecraft fit together.

It should be noted that these spectral plots have not been mathematically computed. The plots are approximations and are used here as illustration. Figure 4 depicts the energy spectrum of an individual up-link; that is, a carrier of frequency f_o phase-modulated by a pseudo-random ranging code plus a voice and an up-data subcarrier. It can be seen that the first order products for the voice and up-data subcarriers appear 30 and 70 kilocycles away from the central carrier spike, respectively. The range code spectral envelope has the sin squared of X over X squared shape with nulls every megacycle. The envelope is actually filled with fine structure defined by the lengths of the various sub-codes making up the range code. This fine structure is neglected for clarity. Figure 5 represents a dual up-link, with two of the individual up-link spectra separated in frequency. The frequency separation is determined by the separation of the two spacecraft phase-modulated carriers and the 240/221 frequency turnaround ratio of the spacecraft transponder. The spectra are of identical form, although the information content of each may be different.

Figure 6 shows the CSM down-link having a phase-modulated carrier at frequency f_{o_1} and a frequency-modulated carrier at frequency f_{o_2}. The frequency separation between the two carriers is kept to 15 megacycles, minimum, due to spacecraft circulator design. The television envelope is always associated with the FM carrier. The range code envelope is always associated with the PM carrier. The two subcarriers bearing real-time voice and telemetry information may be associated with the PM carrier and the two subcarriers bearing recorded voice and telemetry information may be associated with the FM carrier, or vice versa.

Figure 6 shows the LEM down-link spectra. Since no requirement dictates that the LEM transmit simultaneous FM and PM modulation, time sharing of the carrier by the ranging code and television is used. It should be noted, however, that television is transmitted only when convenient and never at the expense of ranging.

Since the LEM carries no tape recorder only one set of subcarriers for voice and telemetry is used. As figure 6 shows, the subcarriers may be transmitted simultaneously with either the ranging code or the television.

Figure 7 shows the dual down-link spectra. The upper spectrum in the figure shows the LEM transmitting a coherent carrier with PM modulation, and the CSM transmitting both the coherent and non-coherent carriers with PM and FM modulation, respectively. The lower spectrum of figure 7 shows the LEM transmitting a non-coherent carrier with frequency modulation, and the CSM transmitting both carriers as in the upper spectrum. The CSM spectra of figure 7 indicate simultaneous transmission of four subcarriers. When this mode of transmission is used, two of the subcarriers are modulated by real-time voice and telemetry information and the remaining two are modulated by recorded voice and telemetry information. However, when recorded information is not transmitted, the real-time voice and telemetry subcarriers may be associated with either the coherent or non-coherent carrier of the CSM spectra.

A merit of the four subcarrier configuration is that for the majority of time when recorded data is not transmitted, the two subcarriers associated with recorded data are available to satisfy any future additional data communication requirements.

Figure 8 shows an individual down-emergency voice link, with base band voice narrow-phase modulated onto the carrier.

Figure 9 shows an individual down emergency key link, with the keyed subcarrier narrow-phase modulated onto the carrier.

3.3 Spacecraft Subsystem Configuration

The previous sections of this report outlined the basic ground and spacecraft systems configurations as well as the type of signals and information to be transmitted to and from the spacecraft. This section will be concerned with the detailed description of the spacecraft equipment used for the performance of the telecommunication and tracking functions.

The LEM S-band subsystem is very similar to that of the CSM in a systems sense. However, as previously pointed out, the LEM does not transmit recorded voice or recorded telemetry. Also, the LEM uses one carrier frequency for transmission to the ground, time sharing FM and PM modulation. The CSM, however, uses two independent carriers, one for FM and the other for PM modulation. Since these carriers may be transmitted simultaneously, recorded voice and telemetry may be transmitted to the ground along with real-time voice and telemetry, and the other functions described in section 3.2.2. The following description is for the CSM; however, a similar description holds for the LEM.

The spacecraft subsystems used for S-band telecommunications and tracking are divided functionally as follows:

1. Premodulation processor

2. Transponder

3. Power amplifier

4. Antennas

These subsystems are discussed individually in some detail. In addition, block diagrams are given to aid the reader in understanding the operation of these subsystem elements. Figure 10 shows the complete spacecraft S-band subsystem.

3.3.1 Premodulation Processor

The premodulation processor is a signal processor which functions to accomplish the following: (1) signal modulation and signal mixing of the information to be transmitted from the spacecraft-to-ground, with the exception of the ranging code, and (2) demodulation of the up-link voice and up-data. This subsystem element consists of a number of filters, subcarrier oscillators, mixing networks, and a number of switches used for selection of the proper modulation technique for any given mode. The block diagram shown in figure 11 outlines the structure of this subsystem. As shown, the premodulation processor has two outputs. These outputs are for frequency and phase modulation, respectively, and are routed to the spacecraft transponder. The PM output can be broken down into four different signals. These signals which can only be transmitted individually, are 51.2 kilobits per second (kbs) or 1.6 kbs real-time telemetry plus normal voice and biomedical data; or recorded PCM telemetry (at 51.2 or 1.6 kbs) plus recorded voice; or emergency voice, or emergency key. The FM output can be broken down into two different signals which can only be transmitted individually, and are television plus real-time telemetry (51.2 or 1.6 kbs) plus normal voice and biomedical data; or television plus recorded telemetry (51.2 or 1.6 kbs) plus recorded voice.

The various types of information which are combined in the mixing networks to produce the various premodulation processor outputs are derived from various sources which are indicated on the left hand side of figure 11. The processing of each of these different types of information is discussed individually.

3.3.1.1 Telemetry (PCM).- Pulse code modulation (non-return to zero) data, having either 51.2 kbs or 1.6 kbs rate, is derived from the spacecraft PCM equipment and may either be recorded or transmitted in real time. Whether played back from the recorder or transmitted in real time, the telemetry information is processed by one of the following methods.

3.3.1.1.1 PM. The PCM data is routed to the bi-phase modulator where the bi-phase modulation is accomplished. That is, the ones and zeros of the PCM data are digitally combined with a 1.024-mc square wave subcarrier. The output of the bi-phase modulator goes through a bandpass filter and is routed to the PM mixing network where it may be summed with other data for modulation on the main carrier and transmission to earth. In this case the summation of the PCM subcarrier and the other data is transmitted to earth via a PM modulated carrier.

3.3.1.1.2 FM. The PCM data is again routed to the bi-phase modulator where modulation is accomplished. Then, the 1.024-mc subcarrier is routed to the FM mixing network where it may be summed with one or with a combination of other signals (such as voice, TV, et cetera). The composite waveform is then routed to the S-band transponder and subsequently transmitted to the ground on an FM carrier.

3.3.1.1.3 Recorded telemetry. During the time period that real time telemetry transmission is not possible (while the spacecraft is behind the moon, for example) the premodulation processor provides switching of the PCM data to the data storage equipment where the data can be recorded. When transmission becomes possible, that data can be played back through the data storage equipment input shown in figure 11, and processed through the recorded telemetry channel using either PM or FM, as discussed previously. As the figure shows, the real time and recorded PCM data can not be transmitted through the same channel. The data can, however, be transmitted simultaneously, one set on FM and the other set on PM.

The real time and recorded telemetry subcarriers have the same frequency; both are derived from the 512 kc PCM clock equipment by doubling the clock frequency. This is a convenient and less complicated way (from an equipment standpoint) to produce a reliable subcarrier.

3.3.1.2 Voice.- As shown in figure 11, there are two different channels in the premodulation processor which are used for the processing of voice signals. One channel accepts voice signals from either the spacecraft audio center or the VHF/AM receiver (which receives voice signals from the astronaut's spacesuit transceiver). The other channel accepts voice signals from the data-storage equipment where voice is recorded during such periods when real-time voice transmission is not possible. The channel which is used for transmission of real-time voice operates in a party line fashion whenever one astronaut is outside the spacecraft, the other in the spacecraft and both wish to communicate with the ground stations. This party line channel employs a voice operated transmission (VOX) push-to-talk (PTT) mechanism which is operated by the astronaut in the spacecraft. Thus, the astronaut who is inside the spacecraft may override the voice of the astronaut who is outside, although both may be listening to voice signals from the ground station. Upon selection of the source, the voice information is routed to the VOX-NO-VOX control, then to a clipper and mixing network, and finally modulated on the 1.25 mc voice subcarrier. The subcarrier with the voice information (and, at times, biomedical data), is then routed to either the FM or PM mixing network and finally to the transponder for transmission to earth.

The recorded voice information, when transmitted, is obtained from the tape recorder, routed to a filter, and modulated on a 1.25 mc voltage controlled oscillator. This modulated subcarrier is then routed to either the FM or PM mixing network and finally, to the transponder for transmission to earth. It should be noted that the real-time voice and recorded-voice subcarriers are not transmitted simultaneously through the FM or the PM channel. As the figure shows, the switch prior to the mixing networks allows transmission of real-time voice via the PM channel and recorded voice via the FM channel, or vice versa.

3.3.1.3 Biomedical data.- During extra-vehicular activities of the astronauts (such as walking on the surface of the moon) some biomedical information will be required by the ground station. This information will be transmitted from the astronaut's spacesuit, via the suit VHF-AM transceiver, and received on the spacecraft transceiver. In turn, the output of the spacecraft transceiver, consisting of the composite of seven subcarriers with their biomedical information, is routed to a filter (high-pass) and then to a mixing network where it is summed with voice information derived from one of the sources discussed previously. The composite output modulates the 1.25-mc subcarrier which is then routed to the FM or PM mixing networks of the processor.

3.3.1.4 Television.- The television information is derived from the onboard television equipment and is routed to the FM mixing network. There, the television information is summed with voice or telemetry, or both, and the composite signal is routed to the transponder for transmission.

3.3.1.5 Emergency voice.- In case of a failure in the spacecraft power amplifier or high-gain antenna, the effective RF power output decreases significantly. This action results in an emergency situation so far as communications are concerned. Under this condition an emergency voice mode is used. In this mode, voice from one of the three sources described previously, is chosen, and the information is routed to the VOX-NO-VOX mechanism, the clipper, and then directly to the transponder phase modulator, completely by-passing the mixing network in the premodulation processor.

3.3.1.6 Emergency key.- In the event of failure of the voice-transmission channel, an emergency key mode will be used. The key mode employs a hand-keyed Morse code signal from the audio center and 512-kc square wave from the central timing equipment (CTE). The code and the 512-kc signal are routed to an "and" circuit through a band-pass filter, and then directly to the transponder-phase modulator, bypassing the rest of the premodulation processor circuitry. The emergency key will appear as a keyed 512-kc subcarrier on the down-link spectrum.

3.3.1.7 Data storage equipment.- Although the data storage equipment (DSE) is not a part of the premodulation processor, this equipment is described very briefly in this section, since DSE outputs are directly routed to the premodulation processor under some circumstances.

The data storage equipment is used for recording of voice and PCM telemetry. Storage of one or both takes place when normal transmission is not possible; for example, when the spacecraft is behind the moon or when the spacecraft is employing a transmission mode which can not be immediately interrupted for transmission of other information.

The onboard DSE has the capability to record 51.2 or 1.6 kbs telemetry and about $4\frac{1}{2}$ hours of voice information. When playback occurs, a patch panel allows the analog voice to be played into the normal voice transmission channel; that is, the frequency modulated subcarrier oscillator (f = 1.25 mc). When the PCM information is played back, the data stream from the DSE is introduced to the bi-phase modulator where it is processed identically as in the normal transmission of PCM telemetry.

3.3.1.8 Up-data and voice.- In addition to processing the information to be transmitted from the spacecraft to ground, the premodulation processor performs the demodulation of the up-data and up-voice subcarriers. As shown in figure 11, there are two demodulators which accept inputs from the transponder and recover the up-link channels.

3.3.2 Transponder

The Apollo spacecraft transponder consists of three basic parts, namely the receiver and two transmitter exciters. This subsystem accepts all outputs from the premodulation processor, performs the final modulation process, and transmits the information to the ground. In addition, the transponder receives all of the information from the ground systems, recovers and turns around the range code, and routes the voice and up-data subcarriers to the premodulation processor. Thus, as far as the unified S-band system is concerned, the transponder performs the transmission and reception functions onboard the command and service module and the lunar excursion module. The unified S-band transponder is shown in figure 12. It is necessary at this point to discuss the operation of the transponder with respect to transmission and reception of information in the various modes discussed elsewhere in this report.

3.3.2.1 <u>Reception</u>.- The ground to spacecraft signal contains its full baseband information, having range code, up-data subcarrier, and voice subcarrier phase modulated onto the carrier. The RF signal enters the transponder at the preselector and is routed to a balanced mixer. There it is mixed with the output of the carrier loop VCO which has been filtered and frequency multiplied by 54. The output of the balanced mixer is then routed to the first IF amplifier. In the IF channel the amplitude of the signal is controlled by the automatic gain control (AGC) signal which is derived from the AGC detector as shown in figure 12. The output of the first IF is routed to the second detector where it is mixed with the output of the carrier loop VCO which has been frequency-multiplied by two. As shown, the second detector has two outputs. One of these outputs is routed to an amplifier and filter (second IF). The other feeds two channels. The first channel consists of an AGC detector which feeds an AGC amplifier filter which, in turn, controls the gain of the first IF amplifier. The second channel consists of a hard limiter which drives the carrier tracking phase detector. The output of the carrier tracking phase detector is bandwidth limited by the IF filter. This output is fed through a filter to the control point on the VCO, thus causing the oscillator to maintain phase coherence with the incoming signal.

As figure 12 shows, there is a second output obtained from the second detector, in addition to the one described above, which is routed to a second IF limiting amplifier and then a wideband detector. The other input to the detector is obtained by dividing the carrier loop VCO frequency by two. The wideband detector provides two outputs which are identical. One of the outputs is routed to a low pass filter, labled "filter and amplifier," where the voice and up-data subcarriers are separated from the range code. The output of this filter is then routed to the premodulation processor for demodulation of the two subcarriers. The other output of the wideband detector is routed to the PM transmitter exciter where it is modulated on the down link carrier and transmitted to the ground.

3.3.2.2 <u>Transmission</u>.- As shown in figure 12, the transponder transmitter has been broken down into the PM exciter and the FM exciter. As the names indicate, the PM exciter is used for transmission in the phase-modulation modes and the FM exciter is used for transmission in the frequency-modulation modes. The PM exciter derives its down-link carrier from the transponder carrier tracking VCO. This is done so that the ground transmitted and received carriers are phase coherent. This phase coherence is necessary for maintaining two-way Doppler track at the ground station. The VCO output, mentioned previously, is denoted in the figure as f_o and is routed to the first phase modulator of the PM exciter, through an RF gate and an X4 multiplier. As shown, the other input to the phase modulator is the output of the wideband detector, described in the previous section. The range code and the up-data and voice subcarriers enter the phase modulator through a video gate, and are phase modulated onto the carrier $(X4f_o)$. The output of the first-phase modulator is then routed to a second-phase modulator where the carrier $(X4f_o)$ is further modulated by the information derived from the PM mixing network of the premodulation processor. Under normal transmission conditions this information consists of voice or PCM telemetry, or both. For non-nominal transmission conditions the input to the second phase modulator consists of either voice information or the emergency key signal, which have been discussed previously. In any case, the output of the second phase modulator is routed to an amplifier, a multiplier chain, an isolator, and an output filter. The output of this filter, as explained in section 3.3.2.3 provides one of the inputs to the S-band power amplifier subsystem for transmission to the ground. Unlike the PM exciter, the FM exciter derives its carrier from a self-contained L-C oscillator. The oscillator signal is frequency modulated by the output information of the FM mixing network of the premodulation processor. This information has been described in section 3.3.1. After modulation, the output of the L-C oscillator is routed to a buffer amplifier and a mixer. Also routed to the mixer is the output of a crystal oscillator which is the basic frequency reference of the FM transmitter exciter. The output of the mixer is routed to an amplifier, a power amplifier, a chain of varactor multipliers, a load isolator, and an output filter. The output of this filter provides the second input to the S-band power amplifier circulator arrangement, which will be explained in the next section. The reader should pay special attention to the fact that the FM and PM carrier frequencies for the command and service module are different from those of the lunar excursion module. This has been explained in the signal design section of this report. For this reason the carrier loop VCO frequency and the L-C oscillator frequency of the command and service module will be different from those of the lunar excursion module. Other than these differences, the operation and design philosophy of the two transponders are the same.

3.3.2.3 S-band power amplifier.- As discussed in the previous section, the transponder performs the transmission and reception functions for the spacecraft. Therefore, failure of this subsystem element results in loss of communications and, consequently, lack of ground knowledge of the location of the spacecraft and physical condition of the astronauts. For this reason redundancy in the transponder is mandatory. Additionally, since simultaneous PM and FM transmission from the spacecraft is used, as discussed elsewhere in this report, a power amplifier circulator arrangement has been devised which will enable the spacecraft to transmit both FM and PM signals at the same time. The power amplifier circulator unit, along with a transponder and its redundant unit, is shown in figure 13. The reader will note from this figure that the power amplifier contains a redundant unit. The redundant unit can be activated by a switch in case of failure in the primary unit or when high power FM and PM signals are transmitted simultaneously.

As shown, the PM output of the transponder may be routed directly to the circulator without entering the power amplifier. This routing is possible only when low power modes are transmitted via the PM channel. If high power PM modes are required, then the PM signal can be routed to the power amplifier; in all cases, the FM output of the transponder must go through the power amplifier, since all FM modes require high power. It should be noted that the power amplifiers used are not linear. For this reason it may not be practical to sum the FM and PM outputs of the transponder prior to, and transmit the composite signal through, the same power amplifier. Because of the importance of the information transmitted via the PM channel of the system, this channel has priority over the FM channel. Therefore, in case of a failure in one of the power amplifiers it is likely that the high power PM modes will be implemented and transmission of FM information will be terminated. The power amplifiers are traveling wave tubes (TWT) which may be driven to provide two power level outputs. One output is 5 watts and the other 20 watts. The low-power output will be used during mission periods when the communication data load is low. The high-power output will be used for full bandwidth transmission from deep space and lunar distances.

3.3.2.4 Antennas.- At this time there is very little information available on the spacecraft antennas. Therefore, this section will describe the spacecraft antennas as they are specified.

There are two different types of antennas on board the spacecraft which will be used for communications at S-band frequencies. These are a high-gain directional antenna and an omnidirectional antenna. The high-gain antenna will be used for transmission from deep space and lunar distances, while the omnidirectional antenna will be used for communications at near-earth distances as well as a back-up to the high-gain antenna in case of failure. The high-gain directional antenna will have the capability of transmitting and receiving signals of bandwidths up to 200 mc at frequencies between 2.0 gigacycles (gc) and 2.4 gc. This antenna will be right-hand circularly polarized and will have a power handling capability of 20 watts. The high-gain directional antenna will be capable of operating at three beamwidths. The half-power beamwidths and corresponding gains are as follows:

$$40° - 12 \text{ db}$$

$$16° - 20 \text{ db}$$

$$6.5° - 28 \text{ db}$$

The omnidirectional antennas will have a power handling capability of 20 watts and a nominal gain of zero decibels.

3.4 Ground Subsystems

3.4.1 General Configuration

Figure 2 was a highly simplified block diagram of the basic, non-redundant, ground system capable of supporting one spacecraft. The figure provided a useful conceptual introduction to the system. However, the real, physical system is not so easily separable into distinct blocks, since the transmitter, reference channel receiver, and ranging circuitry are rather intricately tied together. Figure 14 breaks down, more exactly, into one reference channel receiver, one angle channel receiver, the transmitter, ranging circuitry, data demodulation and premodulation circuitry, and acquisition and programing circuitry in order to show explicitly the interconnections of the various parts. Since the reference channel receivers are identical, and the angle channel receivers are identical, figure 14 shows the minimum blocks necessary for an understanding of the system operation. It should be recognized that there is much peripheral equipment in the ground station which has not been shown in figure 14. The following treatment of the various ground subsystems will be keyed to figures 14 and 15. Although figure 15 is an expansion of figure 14, it is still very much simplified. Most switching, amplification, attenuation, and like circuitry have been deleted. Only that circuitry which contributes to the understanding of the system function has been retained.

3.4.2 Antennas

There are, in the Apollo ground stations, two types of S-band ground antennas. One type, used for deep-space communication is an 85-foot parabolic antenna. The other, for near-space communication is a 30-foot parabolic antenna. Since, at this time, the 85-foot antennas for Apollo have not been specified, this report will consider only the 30-foot antenna.

3.4.2.1 *Main near-space antenna.-* The main antenna is a 30-foot diameter parabolic reflector with 12-foot focal length, having a Cassegrain feed, supported on an X-Y mount. The antenna is capable of tracking at a maximum rate of 4 degrees per second with maximum acceleration of 5 deg/sec^2 down to 2 degrees above the horizon, excepting a lower limit of 10 degrees in the north and south keyholes, with a pointing accuracy of ±90 seconds of arc. The antenna feed produces monopulse two-dimensional sum and difference information from which the angle-error drive signals for the antenna mount servo-system are derived. The transmitting and receiving feeds are circularly polarized, of opposite sense, manually changeable, with a maximum axial ratio of 1.0 db. The receiving gain for properly polarized waves is 44.0 db above isotropic. The first sidelobes are specified to be down more than 30 db from the maximum of the main lobe. The main beam-width is approximately 1.0 degree.

The antenna and associated drive system are capable of operating in eight different modes. These are manual, slow (manual velocity), programed, slaved (to azimuth-elevation source), spiral scan, acquisition track, automatic track, (includes memory track), and test.

3.4.2.2 *Near-space acquisition antenna.-* It is required that an auxiliary antenna be available to aid the main antenna in initial acquisition and lock-on of the received signal. This acquisition antenna will be rigidly mounted on the periphery of the main antenna. Having an approximate diameter of 3 feet, the acquisition antenna has a maximum gain of 22 db, a beam-width of roughly 10 degrees, and a tracking accuracy of 0.5 degrees, total. The acquisition feed is of the simultaneous lobing type whose output is circularly polarized with polarization sense that can be set manually.

3.4.3 Microwave Circuitry

As there are two kinds of antennas associated with the deep-space and near-earth stations, so, also, there are two kinds of microwave circuitry. Figures 16 and 17 show the circuitry for the deep-space stations and near-earth stations, respectively.

Figure 16 shows the separate sets of antenna feeds for the main antenna and acquisition antenna, feeding polarization selector switches. The switches are arbitrarily shown in the right hand circular polarization (RHCP) position. The main angle signals from the polarization switches are always routed to the main angle channel receivers. Likewise, the acquisition angle signals are always routed to the acquisition channel receivers. The sum channel signals are routed from their respective polarization switches to separate diplexers. The receiver outputs of the two diplexers are routed to a switching network which enables the main sum channel to feed low noise amplifier number one and the acquisition sum channel to feed low-noise amplifier number two, or vice versa. This switching network also provides the capability for driving either low-noise amplifier with an RF calibrated noise source.

The outputs of the two low-noise amplifiers are routed to a reference channel selector network, which enables the main reference channel receiver to be fed by the main-sum channel signal and the acquisition reference channel receiver to be fed by the acquisition-sum channel, or vice versa.

The output of the ground station transmitter is routed to a switching network that enables it to feed either diplexer or a dummy RF load. Thus the transmitted signal may propagate through either the main or acquisition antennas.

Figure 16 describes a station capable of supporting only one spacecraft at a time. Figure 17 shows the microwave circuitry for a near-earth station, capable of supporting one spacecraft. The antenna feed polarization for this type station does not switch automatically, but must be changed manually. The main angle signals are routed directly from the main feeds to their respective main angle channel receivers. The main sum channel feed is connected to a diplexer, whose receiver output is routed through a low-noise amplifier and directional coupler to the main reference channel receiver. It is seen that the station transmitter signal is routed through the diplexer into the main sum channel feeds. The transmitter does not drive the acquisition-sum channel feeds.

The acquisition angle signals are routed from the acquisition feeds through low-noise amplifiers, directly to the respective acquisition angle channel receivers. The acquisition-sum-channel signal is routed from the acquisition feed through a low-noise amplifier to a switch. The other input to the switch is the main-sum channel signal, from the directional coupler. This switch allows the acquisition-reference-channel receiver to be fed by either the acquisition-sum-channel signal or the main-sum-channel signal.

3.4.4 Reference Channel Receiver

The following description is for PM, or phase-locked, modes of operation. When the reference channel is used for wideband FM reception, the receiver phase-locked loop is inactive, and the reference channel is simply a manually tuned double-conversion super-heterodyne receiver.

The reference channel portion of figure 15 is labeled Receiver RF and Automatic Gain Control Circuitry. This portion is shown separately in figure 18 for clarity.

The reference channel has three basic functions. One is to translate the incoming composite signal to an intermediate frequency, and position it in the center of the IF passband, using a phase-locked channel, tracking the central carrier component of the signal. The second function is to provide an automatic gain control voltage and carrier acquisition signal, activated only by the carrier component, not by noise. The third function is to provide a sinusoidal signal, phase coherent with the received carrier component, from which either one-way or two-way Doppler frequency shift may be determined. There are other operational functions, but these three are the basic conceptual functions.

The principal oscillator for the reference channel receiver is the stable voltage controlled oscillator (VCO). The twenty megacycle fixed-frequency reference signal is frequency divided by 2 to provide a 10 megacycle reference and also multiplied by 3 to provide a 60 megacycle reference. The 60 and 10 megacycle references are used to establish IF frequencies, fixed at 50 and 10 megacycles. Additionally, the 10 megacycle reference is used in the RF Doppler detection process in such a way that the frequency stability of the 20 megacycle fixed frequency reference does not affect Doppler accuracy.

The incoming composite signal is fed through the S-band amplifiers and preselector to the first mixer. The other input to the mixer is a suitably frequency-multiplied version of the 23.4-mc VCO signal. The loop locks in such a way that the output of the first mixer has a carrier component at 50 mc. This 50-mc component, when heterodyned in the second mixer by the 60-mc reference signal, is further frequency translated to 10 mc, then introduced to the main carrier phase detector, which is also fed the 10-mc reference signal. It is the error signal produced by this phase detector which causes the main loop to lock up in the above described manner. It is seen that the outputs of the two mixers are intermediate frequency signals at 50 and 10 mc. These two IF sources feed the rest of the reception equipment; the 10-mc signal being used for the PM modes and the 50-mc signal for the FM modes.

3.4.5 Angle Channel Receiver

The discussion of the angle channel receiver is keyed to figure 19. Figure 19 is included for clarity. The angle channel strongly resembles the reference channel down through the phase sensitive detector but here it becomes the angle error detector. The angle error detector is basically a phase detector. When the reference channel carrier-tracking loop is locked, the phase of the receiver VCO signal is locked to the phase of the carrier component provided by the sum channel of the antenna monopulse feed. The VCO provides the signal which is heterodyned in the first mixer with the carrier component provided by the difference (or "error") channel of the monopulse feed.

If the RF axis of the monopulse antenna feed differs in direction from the Poynting vector (propagation direction) of the received signal, the amplitude and phase of the carrier component in the angle channel is a direct measure of this directional difference, or error. The angle error detector provides a d-c error signal with the proper magnitude and polarity to cause the antenna servo system to align the RF axis of the feed with the arriving signal wave.

3.4.6 Transmitter

The transmitter in figures 15 and 20 is that circuitry labeled "Transmitter and Frequency Generating Circuitry." This subsystem includes a basic Rubidium frequency standard, a frequency synthesizer phase-locked to the standard, and a master voltage controlled oscillator. The VCO is phase-locked to the frequency synthesizer and provides the RF driving signal for the transmitter. The synthesizer provides the tuning, or frequency changing, capability for the transmitter. The frequency of the synthesizer is changed manually by the operator in discrete frequency steps. Because the frequency changes in steps, the phase-locked VCO is used as the master transmitter oscillator; since the response of the loop to an input frequency step is a continuous smooth transition of the VCO to a new frequency. This smoothness of frequency and phase change of the transmitter is necessary to insure that carrier phase-lock in the spacecraft receiver is not broken during ground transmitter tuning. The VCO feeds a multiplier chain which feeds the phase modulator. The other inputs to the modulator are the range code, up-data subcarrier, and up-voice subcarrier. From the modulator, the composite signal is routed through suitable multipliers and the power amplifier to the microwave circuitry.

3.4.7 Ranging Circuitry[1]

The ranging circuitry is made up of those parts of figure 15 labeled "Receiver IF Clock Demodulation, Code Correlation Circuitry", "Receiver Coder Clock Transfer Loop and Clock Doppler Detector", "Receiver RF Doppler Detector and Transmitter Clock Generator", and "Digital Ranging Equipment". Figure 21, 22, 23, and 24 show these units separately.

The 10-megacycle IF signal from the reference channel receiver contains, in its phase, the transponded five-element ranging code which has a square-wave clock signal as one of the components. This 10-megacycle signal is introduced to a balanced heterodyne detector. The other input to the detector is the 10-megacycle reference which has been phase modulated by the signal from the receiver code generator (receiver coder). The code produced by the receiver coder can assume any one of seven programable states. As far as the two codes are concerned, the balanced detector digitally combines them using the Boolean "Module Two Addition" (exclusive or) combining function. The output of the balanced detector depends upon the program state of the receiver coder and upon the phase of its output with respect to the transponded code. Table VI lists the program states of the receiver coder. It is noticeable that the clock component, c_1, does not appear in the receiver coder output. This fact is the key to the operation of the circuitry.

Volume II shows that the output of the balanced detector, for program state P7, with all code components in phase (acquired), contains a sinusoidal component at the clock frequency at all times. Also, for states P1 or P2, where no code components have been acquired, there is a clock spectral component present in the output of the balanced detector half the time, on the average. During the intermediate program states, while the transponded code components are being acquired or matched in phase, one component at a time, the output of the balanced detector contains a clock sinusoidal component between 25 percent and 100 percent of the time, on the average.

---

[1] It is suggested that section 4.1.1 be read before reading this section. Section 4.1.1 details the basic operation of the pseudo-random-code ranging system.

The output of the balanced detector is fed through a bandpass limiter, of 2 kc bandwidth, centered at the nominal clock frequency, to a phase-locked loop, with a closed-loop noise bandwidth of the order of a few cps. The purpose of this loop is to regenerate or clean-up the received clock signal. The regenerated pure clock, with phase properly adjusted, is fed to a quadrature phase sensitive detector which is also fed the output of the balanced detector. The purpose of this detector is to produce an indication of the correlation between transponded code and receiver generated code. The correlation indication used is the d-c output of the quadrature phase detector. For state P7, with all receiver generated code components matched in phase to the transponded code components, the d-c output of the quadrature detector will be some maximum value. This value is taken to indicate 100 percent correlation of the receiver generated and transponded codes. For other program states, where the receiver generated code components are being shifted in phase to match or acquire the transponded components, the received clock sinusoidal component will be present less of the time, on the average, and the effective d-c output of the quadrature detector will be less. It is the varying effective d-c output of the detector which is used to indicate varying degrees of correlation between the transponded and receiver generated codes, and which indicates the acquisition of a code component.

The pure regenerated clock signal is also fed to the receiver coder clock transfer loop whose function is to supply a frequency doubled version of the clock to the receiver code generator and a normal clock signal to the clock Doppler detector. A phase-locked loop is required, following the switch shown, so that when the switch is thrown, changing the receiver coder driving signal from the transmitter clock reference to the received clock, the frequency and phase of the coder driving signal will change smoothly and continuously, rather than in an abrupt step. The clock output of the receiver coder clock transfer loop is fed to the clock Doppler detector where it is heterodyned with the transmitter clock reference. The resulting clock Doppler signal is routed to the range tally. The transmitter clock reference and a doubled version of it are routed to, and drive, the transmitter code generator.

The remaining part of the ranging circuitry to be described is the receiver RF Doppler detector and transmitter clock generator, shown in figures 15 and 23. It is required that the transmitter clock signal be phase coherent with the transmitter carrier. Then the receiver transponded clock will be phase coherent with the received carrier and the clock Doppler signal will be phase coherent with the RF Doppler signal. This enables an increase in range resolution to be obtained by switching from counting the clock Doppler frequency to counting the RF Doppler frequency. The transmitter clock is obtained by coherently frequency shifting the 22-mc master transmitter VCO signal.

The basic principle of the RF Doppler detection mechanism is to derive from the transmitted and received carrier components a sine wave signal having a frequency which has a known relationship to the relative velocity of the spacecraft with respect to the ground station. Figure 25 is keyed to the following explanation. The notations on the figure are the various normalized signals. In this description it is important to remember that a frequency multiplier multiplies phase as well as frequency, and a mixer subtracts phase as well as frequency.

There are three basic RF generators in the ground equipment. These are the master transmitter VCO, receiver VCO, and the stable reference oscillator. For this discussion, we will represent the radian frequencies of these generators as ω_t, ω_r, and ω_o, respectively. The explanation is best made by following the RF signal from the transmitter VCO and noting its frequency at each step. The transmitter VCO carrier component, $\cos \omega_t t$, is frequency multiplied by 96 and transmitted to the spacecraft. The spacecraft receives and transmits this signal (with one-way Doppler), coherently frequency shifting it by a factor, 240/221. The signal is received at the ground with its frequency multiplied by the two-way Doppler coefficient, $K_D = 1 - \frac{2V}{C}$ where C is the velocity of light and V is the spacecraft radial velocity which is of positive sense when directed away from the ground station. The ground received signal is mixed first with a signal from the receiver VCO which has been frequency multiplied by 96. The output of the first mixer is then mixed with a signal from the stable reference oscillator which has been frequency multiplied by 5. An arbitrary phase of ϕ_o is taken with the reference signal. The output of the second mixer is then applied to a phase detector whose other input is the reference signal, frequency divided by 2. Doppler measurements can only be made when the ground receiver carrier tracking loop is locked. For the locked condition the significant output of the phase detector is zero. This allows the representation of ω_r, the receiver VCO frequency, in terms of ω_t and ω_o.

Returning now to the master transmitter VCO, the carrier component is frequency multiplied by 3 and fed to a mixer. The other input to the mixer is a frequency multiple of 3 times the receiver VCO signal. The output of the first mixer is fed to a second mixer which is also fed by a signal from the transmitter VCO which has been frequency multiplied by a factor 57/221. The output of the second mixer is frequency multiplied by 8 and fed to the final Doppler detector. The other input to this detector is the reference signal which has been effectively frequency multiplied by a factor 5/8.

There are several fine points to be noted in the preceding process. First, any possible instability or drifting in the frequency, ω_o or phase, ϕ_o, of the reference oscillator has absolutely no effect on the Doppler measurement, since these effects are canceled out in the final Doppler detector. Secondly, the frequency of the final Doppler signal contains the Doppler information in the form of a frequency factor, V/c. It may be shown that for spacecraft radial velocities in the order of 10,000 meters per second, the Doppler detector signal has a frequency of the order of 38 kc.

3.4.8 Demodulation Circuitry

The discussion of the data demodulation equipment is keyed to figures 15 and 26. The two inputs are the 50-mc and 10-mc IF outputs of the reference channel receivers. The 50-mc channel is used during reception of the FM modes while the 10-mc channel is reserved for the PM modes.

Except for television and emergency voice, there are two steps in the data demodulation. The first step is the recovery of the individual subcarriers. The second step is the demodulation of the individual subcarriers. The functional explanation of this circuitry can best be made by treating the FM and PM modes separately.

3.4.8.1 **PM modes.-**

3.4.8.1.1 Voice. The normal (or recorded) voice, which is frequency modulated on a 1.25-mc subcarrier and then phase modulated on the carrier, enters the data demodulation circuitry via the 10-megacycle IF channel and is introduced to the main PM demodulator. The 10-megacycle input signal is synchronously demodulated by a stable 10-megacycle VCO signal, phase-locked to the incoming carrier component. The voice subcarrier is recovered by the loop and appears at 1.25-mc in the video output of the demodulator.

The recovered voice subcarrier is routed to the paralleled inputs of the voice subcarrier demodulator and PCM telemetry subcarrier demodulator. A bandpass filter passes the voice subcarrier into the normal voice demodulator which is a modulation tracking phase-locked loop. The output of the loop, which is the baseband voice signal is low-pass filtered to pass either the realtime or recorded voice signal to the voice output.

3.4.8.1.2 Extra-vehicular astronaut (EVA) voice and biomedical telemetry. At the spacecraft the EVA voice and seven frequency modulated biomedical subcarriers were summed with the normal spacecraft voice signal, frequency modulated on the 1.25-mc subcarrier and then phase modulated on the carrier. This composite signal enters the data demodulation circuitry via the 10-megacycle IF channel. The modulated subcarriers are recovered along with the voice baseband, as in 3.4.8.1.1. The subcarriers are then bandpass filtered to individual subcarrier demodulators. The seven information channels are routed to the data handling equipment. The EVA voice is low-pass filtered to the voice output.

3.4.8.1.3 PCM telemetry. The PCM telemetry (realtime or recorded), which was phase-shift keyed on a 1.024-mc subcarrier and then phase modulated on the carrier, enters the data demodulation circuitry via the 10-megacycle IF channel and is introduced to the main PM demodulator. The subcarrier is recovered through synchronous detection and appears in the video output at 1.024 mc. The recovered telemetry subcarrier is routed to the paralleled inputs to the normal voice demodulator and PCM telemetry demodulator. A bandpass filter passes the telemetry subcarrier in to the telemetry detector. The bi-phase modulated subcarrier enters two parallel channels. The first contains a times 2 frequency multiplier. The output of the multiplier is a stable spectral line or sinusoid at twice the subcarrier frequency, or 2.048 mc. In multiplying by 2, the bi-phase modulation is effectively removed. A phase-locked loop is next used to clean up the 2.048-mc signal. The VCO signal, at 2.048 mc, is then frequency divided by two to obtain a stable reference signal at the subcarrier frequency, 1.024 mc. The reference signal is then routed to the same mixer into which the modulated subcarrier was introduced. The output of the mixer is the PCM bit stream and noise, which is routed to the bit synchronizer and then to the data handling equipment.

3.4.8.1.4 Emergency voice. The emergency voice signal, which was phase modulated directly on the carrier, enters the data demodulation circuitry through the 10-megacycle IF channel and is introduced to the main PM demodulator. The voice information is recovered through synchronous detection and appears at the output of the main PM demodulator where it is low pass filtered and passed to the data handling equipment.

3.4.8.1.5 Emergency key. The emergency key signal is a manually keyed 512-kc sine wave, phase modulated on the carrier. This signal, which appears as a subcarrier, enters the data demodulation circuitry via the 10-megacycle IF channel and is introduced to the main PM demodulator. The recovered modulation is routed to the emergency key detector which contains both a 512-kc synchronous detector and an envelope detector. The outputs of these detectors are the emergency key outputs.

3.4.8.2 FM modes.-

3.4.8.2.1 Voice. The normal (or recorded) voice, which is frequency modulated on a 1.25-mc subcarrier and then frequency modulated on the carrier, enters the data demodulation circuitry via the 50-megacycle IF channel and is introduced to the main FM demodulator, a modulation tracking loop. This loop has a closed-loop noise bandwidth which is adjustable to optimize reception of each particular mode. The voice subcarrier is recovered and appears at the output of the modulation tracking loop. The recovered 1.25-mc subcarrier is then routed to the input of a voice demodulator where it is treated in the same manner as in the PM mode.

3.4.8.2.2 Extra-vehicular astronaut (EVA) voice and biomedical telemetry. The biomedical telemetry, summed with normal and EVA voice, enters the data demodulation circuitry via the 50-mc IF as a frequency modulated 1.25-mc subcarrier, frequency modulated on the carrier. The composite signal is routed to the main FM demodulator, which recovers the subcarrier. The modulated 1.25-mc subcarrier is then routed to a voice demodulator. The recovered subcarriers are then treated as in the preceding PM mode.

3.4.8.2.3 PCM telemetry. The PCM telemetry, which was phase-shift keyed on a 1.024-mc subcarrier and then FM modulated onto the carrier, enters the data demodulation circuitry via the 50-mc IF channel and is introduced to the main FM demodulator. The telemetry subcarrier, which is recovered by the main FM demodulator, is routed to a PCM telemetry demodulator where it is treated in the same manner as in the PM mode.

3.4.8.2.4 Television. The television signal, which was frequency modulated directly on the carrier enters the data demodulation circuitry via the 50-mc IF channel and is introduced to the main FM demodulator. The television signal is completely demodulated in this first modulation tracking loop. The baseband television signal is routed through a low-pass filter to the output.

3.4.9 Premodulation Circuitry

The premodulation circuitry, shown in figures 15 and 20, is fed the baseband voice and up-data signals. The voice signal is frequency modulated on a 30-kilocycle subcarrier, the up-data on a 70-kilocycle subcarrier. The outputs of the two subcarrier oscillators are then summed and routed to one modulation input of the transmitter phase modulator. Into the other phase modulator input is fed the transmitter ranging code. In the phase modulator, the three up-going information signals are effectively summed.

3.4.10 Acquisition and Programing Circuitry

The sequential steps in acquisition and ranging are explained in section 4.1.2. The circuitry associated with these steps is indicated in a very simple manner in figures 15 and 24. Although a complete explanation of the functioning of this circuitry is beyond the scope of this report, the physical operation will be described in simple terms.

The acquisition and programing circuitry monitors portions of the ground system and senses the results of various operations. In particular, the d-c outputs of the carrier loop quadrature phase detector and the correlation detector in the range clock receiver indicate to the acquisition and programing circuitry the state of carrier acquisition and range code acquisition, respectively. This circuitry actually controls the shifting of the receiver code and up-dates the range tally while shifting the code. It should be noted that antenna drive programing prior to signal acquisition is not performed by this circuitry, but by a separate subsystem, described in section 3.4.11.

3.4.11 Ground System Peripheral Equipment

The previous material has described those ground subsystems which are a direct part of the S-band system. However, there is much peripheral equipment required in the operational ground system. This section describes briefly some of the peripheral equipment used operationally. It disregards items such as signal test units, control consoles, equipment racks, and so forth, which are also regarded as peripheral equipment. The equipment to be described in this section is as follows:

1. Antenna position programer

2. Tracking data processor

3. Up-Data verification receiver

4. PCM signal conditioner

The equipment listed above has recently been specified by Goddard Space Flight Center.

3.4.11.1 Antenna position programer.- The antenna position programer is a special subsystem in the ground station. Its function is to accept spacecraft predictional data and define spacecraft position in the visible sky region of the ground station in question. The predictional data are transmitted to the remote site from the main control center of the S-band ground network.

The processor uses a general purpose computer which accepts the predictional data and corrects these data for errors caused by transmission from the control center to the remote site. Then the corrected data are used to generate antenna drive tapes with words containing time and X, Y, command angles. These parameters are generated every second or every ten seconds. Additionally, the programer reads the real X and Y angles from the antenna position encoding subsystem at a rate of 10 per second. The programer compares the real parameters with the predictional parameters, and, if a difference exists, generates analog voltages at the rate of 10 per second to correct the antenna control and drive system. Thus, the antenna is provided with a backup mode (program mode) should the primary or auto-track mode become inoperable.

3.4.11.2 *Tracking data processor*.- The tracking data processor is a subsystem which consists of a computer, data storage units, teletype equipment, Doppler counters, tape recorders, and a number of gating networks and controls. This subsystem provides time, X and Y angle information, and range and Doppler information in serial form, which is compatible with two types of ground communication links; the high speed (2,000 bits per second), and the low speed (60 words per minute) teletype link. The processor arranges the above data in a proper format and, at the same time, provides station identification and other functional information, as required by the network control center. The inputs to the processor are derived from the ranging subsystem, the antenna shaft angle encoding subsystem, the timing subsystem, and the tracking receiver. In addition to the high and low speed data, the processor provides recording of all data on a magnetic tape recorder. Then, in case of a communications failure between remote site and the control center, the recorded data can be processed and analyzed at a later time. Provisions are also made for the conversion of the slow speed data from binary to decimal form and for printing of these data in clear text, paged form.

3.4.11.3 *Up-data verification receiver*.- The up-data verification receiver is a subsystem which is used to determine whether up-data and voice have been transmitted correctly from the ground station to the spacecraft. This subsystem consists of a phase locked device for carrier demodulation and 30- and 70-kc discriminators for subcarrier demodulation. The input to the receiver is derived from the output of the power amplifier. Upon demodulation the voice signals are monitored for quality. The demodulated up-data signals are compared with those at the output of the digital command system. If an error is detected, correction procedures are immediately initiated. In addition to the preceding comparison, the transmitted voice and up-data information is recorded. Thus, a record of the up-link transmitted information exists at all times.

3.4.11.4 *PCM signal conditioner*.- The PCM signal conditioner subsystem is used to test and analyze the performance of the spacecraft-to-ground telemetry channel. This subsystem generates a digital pulse train at the "PCM signal simulator" with bit rates of 10 to 200 kbs. This signal is mixed with specific quantities of noise and is capable of driving the up-data subcarrier modulator of the ground system. Additionally, this subsystem is capable of providing pure signals (without noise) to the PCM bit comparator, or a mixed signal and noise to the bit synchronizer and signal conditioner, and a clean signal (no noise) to the bit comparator. The bit synchronizer and signal conditioner accept signals from either the PCM simulator, described above, or the output of the ground telemetry demodulator. The signal conditioner accepts PCM signals under adverse signal-to-noise conditions and reconstructs these signals to their optimum shape with the least noise possible. Finally, the bit comparator may receive signals from the signal simulator and the signal conditioner. These two signals will be made coincident in time. Further, the bit comparator can compare the two signals and route an error signal to the frequency counter when the inputs in question are not of the same time, phase, or polarity.

4.0 SYSTEM OPERATIONAL TECHNIQUES

4.1 Ranging and Spacecraft Acquisition

4.1.1 A Physical Explanation of the Ranging Process

Most of the modulation and information processing techniques employed in the unified system are relatively standard and well documented. These techniques include the use of PCM telemetry and analog voice on subcarriers. However, since the JPL pseudo-random code ranging technique employs a system mechanization that is complex, a general physical explanation of this process is required. As with most ranging schemes, this one depends on measuring the round-trip propagation time for some event which is generated by the ranging station. The uniqueness of this technique lies in the nature of the event and the measurement mechanization. Figures 27 and 28 provide insight into the following explanation.

The event generated by the ground station is a periodic binary waveform. This waveform is composed of a repeating binary sequence, having a bit length of one microsecond and an unambiguous length of several million bit periods. This length insures that the periodic waveform, or code, does not repeat during the round trip electromagnetic wave propagation time from ground to spacecraft.

The heart of the basic system consists of two code generators which produce the two range codes described in previous sections of this report, and a range tally, which is a digital machine whose function is to generate range measurement data in response to certain digital inputs. One code generator is driven by a stable clock generator. This generator is the Transmitter Range Code Generator. The other generator, the Receiver Range Code Generator, is driven by a separate clock generator which is not fixed in rate but, rather, may be phase-locked to a clock signal at its input. Additionally, the receiver generator has the capability to change the phase, or time-delay, of its code in integral bit length steps, in response to digital instructions. Supplementary parts of the basic system are a cross-correlator and indicator to provide a means of knowing when a receiver-generated code is in phase or correlated with the incoming transponded code; and a Doppler frequency detector and counter which indicates to the range tally the cumulative phase differences between the transmitter and receiver clock generators.

The range measurement is made as follows: initially the receiver code generator is synchronized in phase to the transmitter code generator and the receiver clock generator is locked to the transmitter clock generator. At this time, the range tally is set to indicate zero range. Meanwhile, the transmitted code is allowed to propagate to the spacecraft and back to the ground. Because of the range and velocity of the spacecraft with respect to the ground station, the code, upon return to the ground, will have a time delay and different clock rate, compared to the transmitted code. Next, the received code generator is freed from synchronization with the transmitter code generator and the receiver clock generator is switched from the transmitter clock generator so that it may acquire and lock to the clock component of the incoming transponded code. At this point, it is helpful to visualize the relationship between the transmitted code and the transponded code as it is received at the ground (fig. 28).

Part 1 of figure 28 shows the range code as it is transmitted from the ground station. A certain arbitrary element of the code is identified at time, t_o. Part 2 shows the transponded range code as it is received at the ground station. At time t_1 the same code element identified at the ground transmitter is received on the ground. The time delay, $\tau_1 = t_1 - t_o$ is a measure of the spacecraft range at a time $t_R = t_1 - \frac{\tau_1}{2}$ since range, R, is given as $R = \frac{C\tau}{2}$ (neglecting Earth rotation) where C is the velocity of light and τ is the two-way propagation time to the spacecraft. In general, because of the relative movement between ground station and spacecraft, the delay time, τ_2, of the code received at time t_2 will not be the same as τ_1. It should be noticed that the received code does not run at the same rate as the transmitted code (a Doppler effect) and is either stretched or compressed in time.

Part 3 of figure 28 shows the receiver-generated code at the start of code acquisition, slaved in rate and phase of the transmitted code. Parts 3, 4, and 5 are conceptually correct, but not quite physically correct, since the receiver-generated code actually lacks the clock element which is a part of the transmitted code. In part 4, at time, t_g, the receiver clock generator is switched away from the transmitter clock generator, to the range clock receiver, and is allowed to acquire the received clock signal. At some later time, t_c, clock acquisition is complete and the receiver generated code is running at the same rate as, and synchronized on a bit to bit basis with, the received code. The receiver generated code is not yet matched in phase, or time delay, to the received code.

Starting at time, t_s, the time at which the receiver generated code starts separating from the transmitted code in phase, it is required to accumulate, or keep track of, the amount of phase change between the receiver generated and transmitted codes. This is done by detecting the clock Doppler, or difference frequency, between the two clock generators and counting or integrating this frequency. Every time the clock difference frequency makes a zero-crossing, the receiver generated and transmitted codes have slipped past each other by one bit period. A count of clock Doppler zero crossings, together with an increasing or decreasing sense indication, effectively keeps track of the cumulative phase difference between the transmitted and received generated codes, caused by their running at different rates. This integrated clock Doppler signal is routed to the range tally to keep the continuous range measurement accurate.

Part 5 of the figure shows the result of the final step in acquisition of the code, that of stepping the receiver code generator in discrete bit steps until it exactly matches in phase, or correlates, with the received code.

This correlation is actually obtained by stepping each of the four receiver code components separately, with the range tally being advised of the equivalent movement of the entire receiver code. After obtaining correlation of the entire code, the continuously up-dated range measurement being displayed by the range tally is correct and has a resolution of about half a bit period or 150 meters. After the acquisition of the transponded code, a step is taken to refine the resolution of the continuous measurement. This is to switch from counting Doppler on the two clock generators to counting Doppler on two RF signals derived phase coherently from the two carriers. Since the clock signals are, themselves, coherent with their respective carriers, the clock Doppler is phase coherent with the RF Doppler. Immediately after tallying a zero crossing of the clock Doppler, the RF Doppler counter is allowed to commence up-dating the range tally. This increases the resolution of the continuous range measurement to about 1.05 meters.

4.1.2 An Explanation of Acquisition

Many signals must be acquired during operation of this system. The following lists the signals and the order in which they must be acquired and gives a brief description of the functioning of the various circuits during acquisition. Many of the acquisition processes are automated, being run by the acquisition and programming circuitry. This explanation does not depend on a knowledge of the exact method employed to automate the functions. Acquisition begins with the spacecraft transmitting a carrier which is phase modulated by the voice and PCM telemetry subcarriers.

At the ground station the acquisition antenna is programmed, using previously determined information, to search for the spacecraft signal. This search may consist of a spiral scan motion combined with programmed tracking of a prediction of the spacecraft trajectory.

Combined with the antenna search in angle, the acquisition reference receiver searches in frequency for the central carrier component of the spacecraft signal. This search is made by sweeping the receiver VCO in frequency. When the VCO frequency is at the proper position relative to the spacecraft carrier component, the acquisition carrier tracking loop will lock. The lock is signified by the appearance of a d-c carrier acquisition signal from the RF channel quadrature phase detector.

Once the carrier tracking loop is locked, the antenna angle drive channels become operative. The antenna drive is then switched from the programmed search to the acquisition-angle channels. When the acquisition antenna is sufficiently alined, the main antenna, which is physically tied to the acquisition antenna, acquires the spacecraft signal. The main reference channel VCO is then swept into lock with the carrier component. This lock activates the main angle channels. The antenna drive is then switched from the acquisition angle channels to the main angle channels.

Next, the carrier tracking loop in the data demodulator is locked up. Then, both the voice and PCM telemetry demodulators are activated. Now, the ground transmitter VCO is swept in frequency until the spacecraft carrier tracking loop acquires. Acquisition is signalled at the spacecraft by a d-c acquisition signal (the spacecraft coherent automatic gain control voltage). The spacecraft in-lock signal and VCO static phase error are telemetered to the ground station.

The ground transmitter frequency is then slowly shifted to adjust the telemetered static phase error indication. This adjustment insures that the signal received at the spacecraft is properly positioned in the spacecraft receiver passband. The voice and up-data subcarriers are next acquired in the spacecraft. When a spacecraft FM mode is being employed, the auxiliary FM ground receive is tuned to acquire the FM signal. Then, the main modulation tracking loop in the data demodulator is locked. Because of the antenna auto-track mechanization, the FM carrier must be acquired prior to FM acquisition. At this point all the communication channels have been acquired. The final step is to perform the ranging. The five-element ranging code, generated by the transmitter range code generator, is allowed to phase modulate the ground transmitter and is sent to the spacecraft. At the spacecraft the code is demodulated and then remodulated on the down carrier. The code is received on the ground and passes through the main reference channel to the receiver IF clock demodulation and code correlation circuitry. Here the clock VCO is swept in frequency and the ranging clock component acquired. Meanwhile, the receiver coder clock transfer loop has been switched and locked to the transmitter clock signal. The receiver code generator has been matched in phase to the transmitter code generator, and the range tally set to zero. The transfer loop is now switched and allowed to acquire the received clock signal, sweeping the coder clock transfer loop VCO, if necessary.

Counting and tallying of clock Doppler begins at this point. The receiver code generator is now stepped in phase until all four of the code elements are acquired, as signalled by the code correlation indication. After acquisition of the complete code, the range tally is switched to count RF Doppler. The transmitter code modulation is then removed. The range tally continues to be up-dated by the RF Doppler, alone. This action completes the acquisition and ranging procedure.

4.2 Lunar Mission Communication Requirements

Detailed chronological communication requirements for all the Apollo missions have not been formulated at this time, but it is possible to outline, in general, the maximum capabilities required for the S-band telecommunication system during the lunar landing mission. This description includes the tacit assumption that the S-band system is the only tracking and communication system in use during the mission. This assumption will not hold during the launch phase, and may not hold during the earth orbital, reentry, and recovery phases. The requirements will be treated according to mission phase.

4.2.1 Prelaunch

Before launch, from the time the CSM systems are activated and after the astronauts have come aboard, the S-band system will be required to transmit voice and telemetry and perhaps television to the Mission Control Center at

Cape Kennedy. It will also be required to check the operation of the system, itself, before launch.

4.2.2 Launch

During the period from the beginning of main booster stage burn, through engine cut-off at insertion into earth orbit, it will be necessary to have S-band tracking, two-way voice, PCM telemetry, up-data, and television. It should be noted that in this and the other phases, the television requirement has the lowest priority. However, given the capability, transmission of television at the indicated times is highly desirable.

4.2.3 Earth Orbit

During the one to four earth orbits before translunar injection, it is required to have tracking from at least two stations on the first orbit and at least one per following orbit. Also required is voice, PCM telemetry, up-data, and television.

4.2.4 Translunar Injection

During the burning of the S-IV-B stage for injection into a translunar trajectory, and depending upon the radio visibility of an S-band ground station, it is desirable to have tracking, voice, and PCM telemetry. During this, as well as the other critical mission phases, if communication is not maintained, voice and PCM telemetry will be recorded on board for later playback. However, it is required that ranging, telemetry, voice, up-data, and recorded data be available before 20 minutes after injection.

4.2.5 Spacecraft Transposition and LEM Checkout

After injection, during the period when the CSM separates from the remaining portion of the vehicle and removes the LEM, ranging, voice, and telemetry are required and television is desirable. Biomedical telemetry may also be required at this time.

During the period when the LEM systems are being checked out, it will be required to have, simultaneously with both vehicles, voice and telemetry. Required at least sequentially with both vehicles will be up-data, biomedical data, television and tracking.

4.2.6 Earth-Lunar Coast

After the LEM checkout period and for the duration of the coast phase, until insertion into lunar orbit, all functions will be required, from one spacecraft only.

4.2.7 Lunar Orbit

During lunar orbit of both vehicles, before separation of the LEM for descent, the LEM systems must again be checked out. It will be required to have tracking, voice, PCM telemetry and up-data from both vehicles, simultaneously, and perhaps EVA biomedical data from one vehicle. Additionally, playback of any PCM telemetry and/or voice recorded by the CSM on the back side of the moon may be required.

4.2.8 Lunar Landing, Surface Operations, and Rendezvous

Immediately upon separation of the LEM for the lunar descent, tracking, voice, PCM telemetry, and up-data will be required for both spacecraft simultaneously. The CSM will continue to require telemetry and voice dump as it orbits the moon. After lunar landing, tracking of the LEM will not be required. However, voice, PCM telemetry, up-data, biomedical data, and television will be required for the LEM during its lunar stay. Simultaneously, the CSM will require all functions except television and biomedical data. For lunar take-off and through rendezvous with the CSM, tracking, voice, and PCM telemetry will be required for both spacecraft simultaneously.

4.2.9 Moon-Earth Injection and Coast

Requirements for this phase are essentially the same as for the analogous earth-moon phases, with the addition of required tracking of the LEM after separation, to establish its lunar orbit ephemeris.

4.2.10 Reentry

Tracking, voice, PCM telemetry, up-data, recorded data playback, and possibly television will be required for all non-blackout times. This again depends on the radio visibility of ground stations.

4.3 System Operation for a Nominal Lunar Mission

The following is a description of communications occurring during a hypothetical lunar mission. It is intended to illustrate the probable sequence of communication events for a nominal mission. The narrative is broken into the logical mission phases.

4.3.1 Prelaunch

Before launch, the spacecraft systems are activated. The launch site S-band tracking station antenna is pointed at the spacecraft and acquires the two command module carriers using the spacecraft omniantenna. The spacecraft then acquires the tracking station carrier. A prelaunch S-band systems checkout operation is then performed. After the preliminary system checkout, the up-link voice and command subcarriers are acquired for operational use. The down-link voice and telemetry subcarriers and television signal are acquired in the FM mode. The range code is transmitted and acquired and range tally of PM carrier Doppler initiated using the high power PM mode.

4.3.2 Launch

During booster burn the launch site tracking station continues to Doppler track the spacecraft in the high power PM mode and receive telemetry, voice and television in the FM mode. As the spacecraft gains altitude it becomes visible to the go-no-go tracking station down-range. This station acquires the spacecraft PM carrier first, then its FM carrier with telemetry, voice and television. At the hand-over time, the launch site station ceases transmitting. The go-no-go station observes this event by monitoring the spacecraft telemetered transponder in-lock signal. Rapidly the go-no-go station commences transmitting and causes the spacecraft to acquire the new carrier, voice, and command channels. The go-no-go station then transmits the ranging code, acquires it, and commences tallying spacecraft PM carrier Doppler.

4.3.3 Earth-Orbit Insertion

For the remainder of powered flight and during insertion, the go-no-go station continues to Doppler track the spacecraft in the high-power PM mode and receive voice, telemetry, and television in the FM mode. When powered flight terminates and after the spacecraft has stabilized, sufficient Doppler track data is accumulated by the go-no-go station to permit the go-no-go decision. After contact with the go-no-go station is lost, spacecraft transmission in the FM mode is terminated. The voice and telemetry channels are switched to the PM mode.

4.3.4 Earth Orbit

For the remainder of time spent in earth orbit, most communication and tracking is performed with the spacecraft in the high-power PM mode, using the omniantenna. The only exceptions are if data has been recorded or if television is required. Then the spacecraft FM mode may be initiated.

4.3.5 Translunar Trajectory Injection

Just before the S-IV-B engine is ignited at the beginning of injection, if contact is not being maintained with a ground station, voice and PCM telemetry will be routed to the onboard data storage equipment. The spacecraft will commence transmitting in the PM mode for ground station acquisition. Depending on the geographical location of the injection point, a ground station may not be acquired until after injection burn is complete. In any event, the two-way acquisition of PM mode carriers, voice, telemetry, up-data, and ranging code will be made in the same manner as before. After acquisition, sufficient tracking data is accumulated to give the injection go-no-go decision.

4.3.6 Transposition and LEM Checkout

After the injection go-no-go decision, tracking and communications continue in the PM mode, using the omniantenna, as the command service module (CSM) separates from the remainder of the vehicle to begin transposition and docking with the Lunar Excursion Module (LEM). Care must be taken to maintain spacecraft attitude during this maneuver so that the PM links will not be broken. Television and/or data playback may be required at this time. If so, the FM mode will be activated. However, to avoid interruption, real-time PCM telemetry and voice will remain in the PM mode. After docking and transposition is complete, an astronaut enters the LEM and activates the LEM systems. From this point of the mission until lunar operations commence, the ground-tracking stations must have dual capabilities in all communication functions except angle tracking and ranging. While operational tracking and communications continue using the CSM PM mode, the LEM PM mode is activated and two-way LEM ground acquisition performed. All LEM communications functions are then checked out. During this period the CSM high-gain antenna is deployed and activated.

4.3.7 Translunar Coast

After the completion of LEM checkout, LEM transmission ceases. For the remainder of the flight to the moon, primary communications with the spacecraft continue, using the CSM PM mode with possible intermittent FM transmission. If an astronaut leaves the spacecraft, his voice and special biomedical data are relayed through the CSM voice channel.

4.3.8 Lunar Orbit

Voice and PCM telemetry will be transmitted in the PM mode during retrofire to insert into lunar orbit. Tracking and ranging will also be performed on the spacecraft, using the PM mode, during retrofire. After insertion into lunar orbit, communication will be accomplished with the CSM in the PM mode. Sufficient tracking data will be accumulated to insure a good orbit around the moon. While the spacecraft is behind the moon, voice and telemetry will be recorded in the data storage equipment. When the spacecraft reappears from behind the moon, two-way signal acquisition is made with the CSM in the PM mode. Ranging and tracking is immediately initiated. Real-time voice and telemetry are transmitted through the PM channel. Recorded telemetry and/or voice are transmitted through the FM channel. For the remainder of the time that the CSM remains in moon orbit, communications and tracking will continue as before.

4.3.9 Lunar Landing, Surface Operations and Rendezvous

At some time during lunar orbit, two astronauts will transfer from the CSM to the LEM. The LEM systems will be activated and a checkout procedure initiated. The CSM will be communicating with the earth station in the PM mode. Two-way PM carrier acquisition will be made between LEM and ground. The LEM telemetry, voice, and up-data channels will be activated, and tracking on LEM and CSM, simultaneously, will be initiated. At this point of the mission full dual capability for all communication functions will be required of the earth tracking stations. After LEM systems checkout has been satisfactorily completed and the landing decision made, the LEM will separate from the CSM and retrofire for descent to the lunar surface. Continuous communication and tracking will be maintained with both spacecraft simultaneously, in their PM modes, visibility permitting. After a satisfactory LEM descent and landing, tracking of the LEM is not necessary, and voice, telemetry, and television from the LEM may be had in the FM mode. During the stay on the moon's surface one astronaut will depart the spacecraft. His voice and biomedical data will be relayed to earth via the LEM voice channel. Meanwhile, the CSM ground link will be maintained in the PM mode, except for possible recorded data playback in the FM mode. After completion of lunar surface operations the LEM will return to the PM mode for two way acquisition of carriers, range code, and information channels. The LEM will then boost free of the surface and establish a rendezvous trajectory toward the CSM. Simultaneous tracking and communication with both spacecraft will again be performed in the PM modes.

4.3.10 Earth Trajectory Injection and Coast

After a successful rendezvous, the two LEM astronauts will transfer back to the CSM. The LEM systems will be left on. The CSM will separate from the LEM and boost into an earth transfer trajectory. Communication and tracking with the CSM will be performed in the PM mode with possible FM television.

As long as the LEM remains in the ground antenna beam-width of the CSM tracking station, or if a second tracking station is visible, the LEM will be tracked in the PM mode to determine its lunar orbit. After injection into the transfer trajectory, the CSM will coast back toward earth with communications the same as during the outward trip.

4.3.11 Atmospheric Reentry

As the CSM nears the earth, communication will be in the PM mode using the omniantenna with possible FM television. Maintenance of communications will depend on the geographical location of the reentry path and the reentry ionization density. Given a visible tracking station, communication in the PM mode will be maintained through landing of the CSM.

5.0 REFERENCES

1. Baumert, L., Easterling, M. Golomb, S. W., and Viterbi, A.: Coding Theory and its Applications to Communications Systems. Technical Report No. 32-67, Jet Propulsion Laboratory, Pasadena, California, March 31, 1961.

2. Martin, Benn D.: The Pioneer IV Lunar Probe. A Minimum Power FM/PM System Design. Technical Report No. 32-215, Jet Propulsion Laboratory, Pasadena, California, March 15, 1962.

6.0 TABLES

TABLE I. - CODE COMPONENT LENGTHS

Code Component	Length (bits)
c_1	2
x	11
a	31
b	63
c	127

TABLE II. - BIOMEDICAL TELEMETRY CHANNEL DATA

Channel	Center Frequency	Channel Information
1	4.0 kc	Spare
2	5.4 kc	SSA Battery
3	6.8 kc	Suit Pressure
4	8.2 kc	Oxygen Pressure
5	9.6 kc	Suit Temperature
6	11.0 kc	Body Temperature
7	12.4 kc	Impedance Pneumograph

TABLE III.- UP-LINK MODULATION MODES (PM)

Mode	Information Transmitted
1-A	Range Code Only
1-B	Voice Only
1-C	Up-Data Only
1-D	Range Code + Voice
1-E	Range Code + Up-Data
1-F	Range Code + Voice + Up-Data
1-G	Voice + Up-Data

TABLE IV.- DOWNLINK MODULATION MODES

MODE	TRANSMITTED INFORMATION		REMARKS
	PM Channel	FM Channel	
A-1	Carrier only		By-pass PM Final Amp.
A-2	Carrier only		
A-3	Carrier only	Television & Real-time PCM T/M, Voice, EVA, Biomed.	By-pass PM Final Amp.
B-1	Range Code		By-pass PM Final Amp.
B-2	Range Code	Television & Real-time PCM T/M, Voice, EVA, Biomed.	By-pass PM Final Amp.
B-3	Range Code & Real-time PCM T/M, Voice, EVA, Biomed.		
B-4	Range Code & Real-time PCM T/M, Voice, EVA, Biomed.	Television & Recorded PCM T/M, Recorded Voice	
C-1	Real-time PCM T/M, Voice, EVA, Biomed.		
C-2	Real-time PCM T/M, Voice, EVA, Biomed.	Recorded PCM T/M, Recorded Voice	
C-3	Recorded PCM T/M, Recorded Voice	Real-time PCM T/M, Voice, EVA, Biomed.	
D-1	Emergency Voice		By-pass PM Final Amp.
D-2	Emergency Voice		
E-1	Emergency Key		By-pass PM Final Amp.

TABLE V.- SUBCARRIER FREQUENCIES

Information Channel	Channel Number	Frequency
Up-link Voice		30 KC
Up-Data		70 KC
Down-link Voice		1.25 MC
PCM Telemetry		1.024 MC
Recorded PCM Telemetry		1.024 MC
Recorded Voice		1.25 MC
Biomedical Data	1	4.0 KC
	2	5.4 KC
	3	6.8 KC
	4	8.2 KC
	5	9.6 KC
	6	11.0 KC
	7	12.4 KC
Emergency Key		512 KC

TABLE VI.- RECEIVER CODER PROGRAM STATES

Program States	Receiver Coder Output
P1	0
P2	0
P3	$\bar{X} \cdot a$
P4	$\bar{X} \cdot a$
P5	$\bar{X} \cdot b$
P6	$\bar{X} \cdot c$
P7	$\bar{X} \cdot \left[(a \cdot b) \vee (b \cdot c) \vee (a \cdot c) \right]$

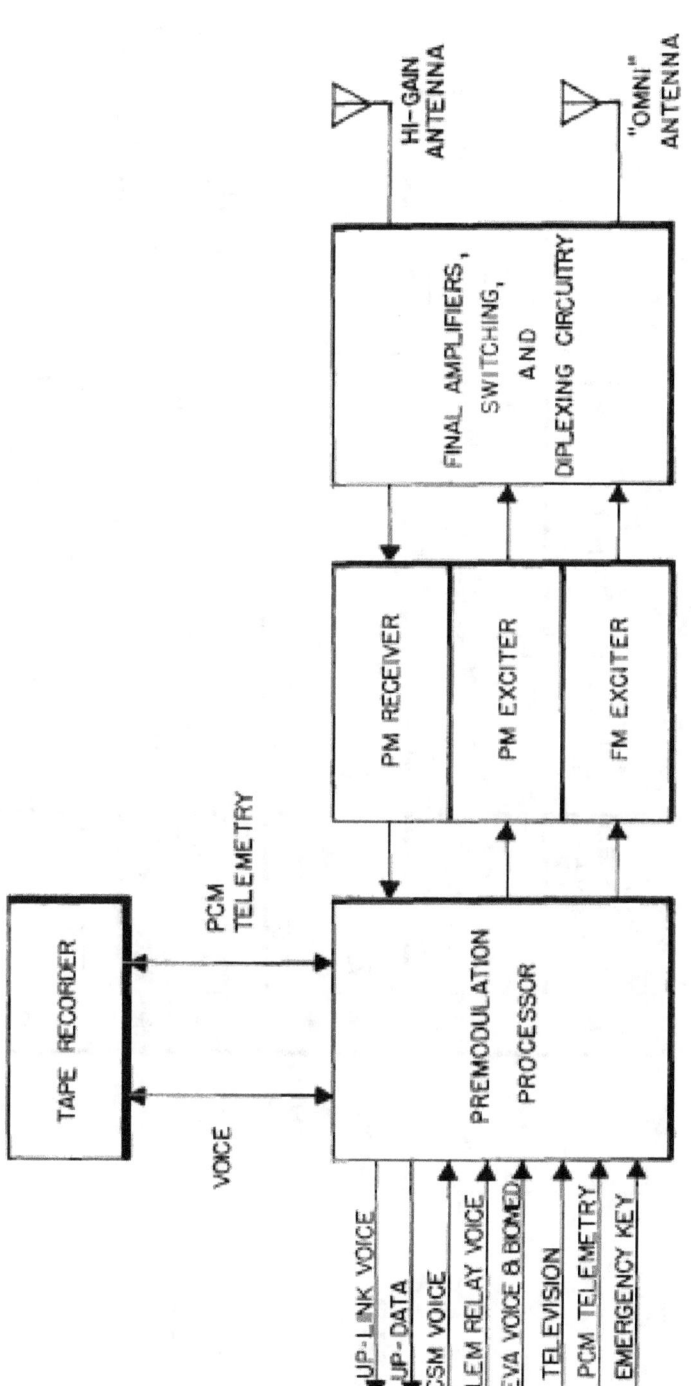

Figure 1.- The basic spacecraft system

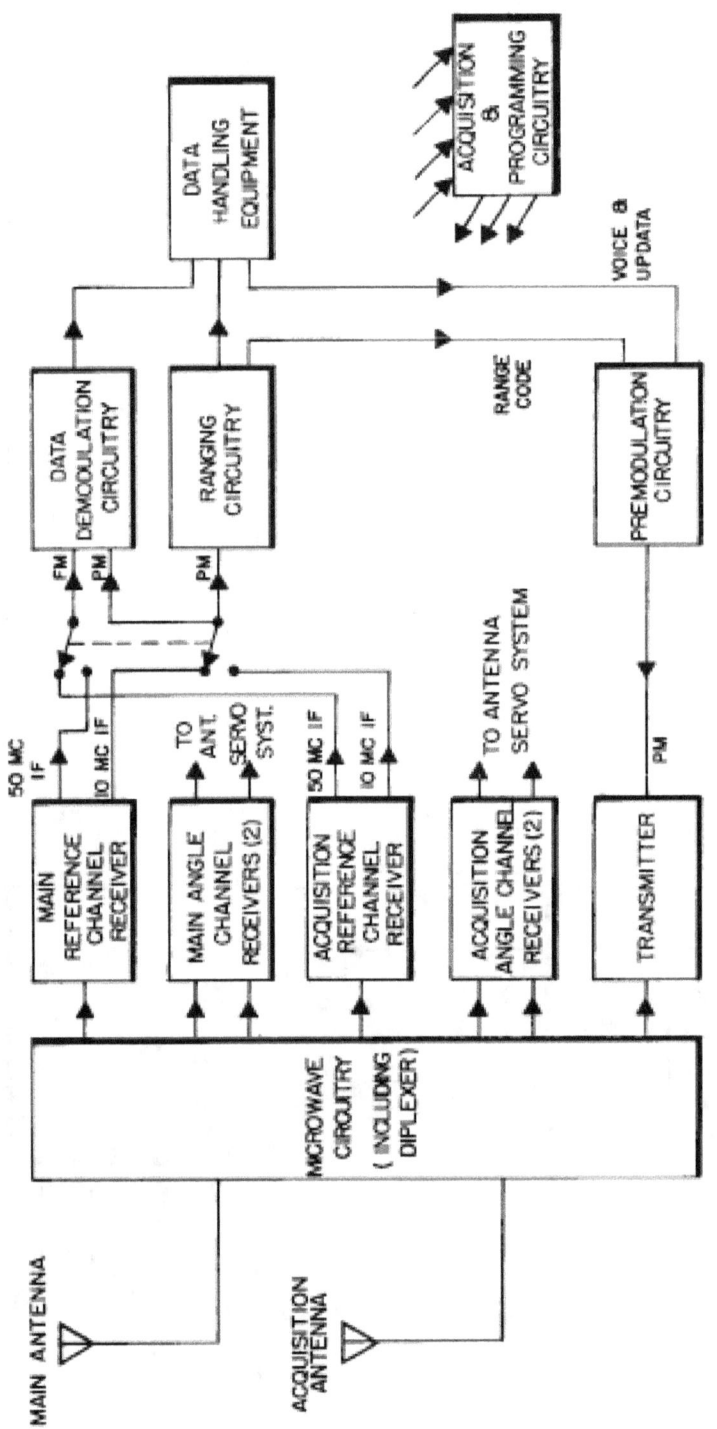

Figure 2.- The Apollo unified S-band ground system

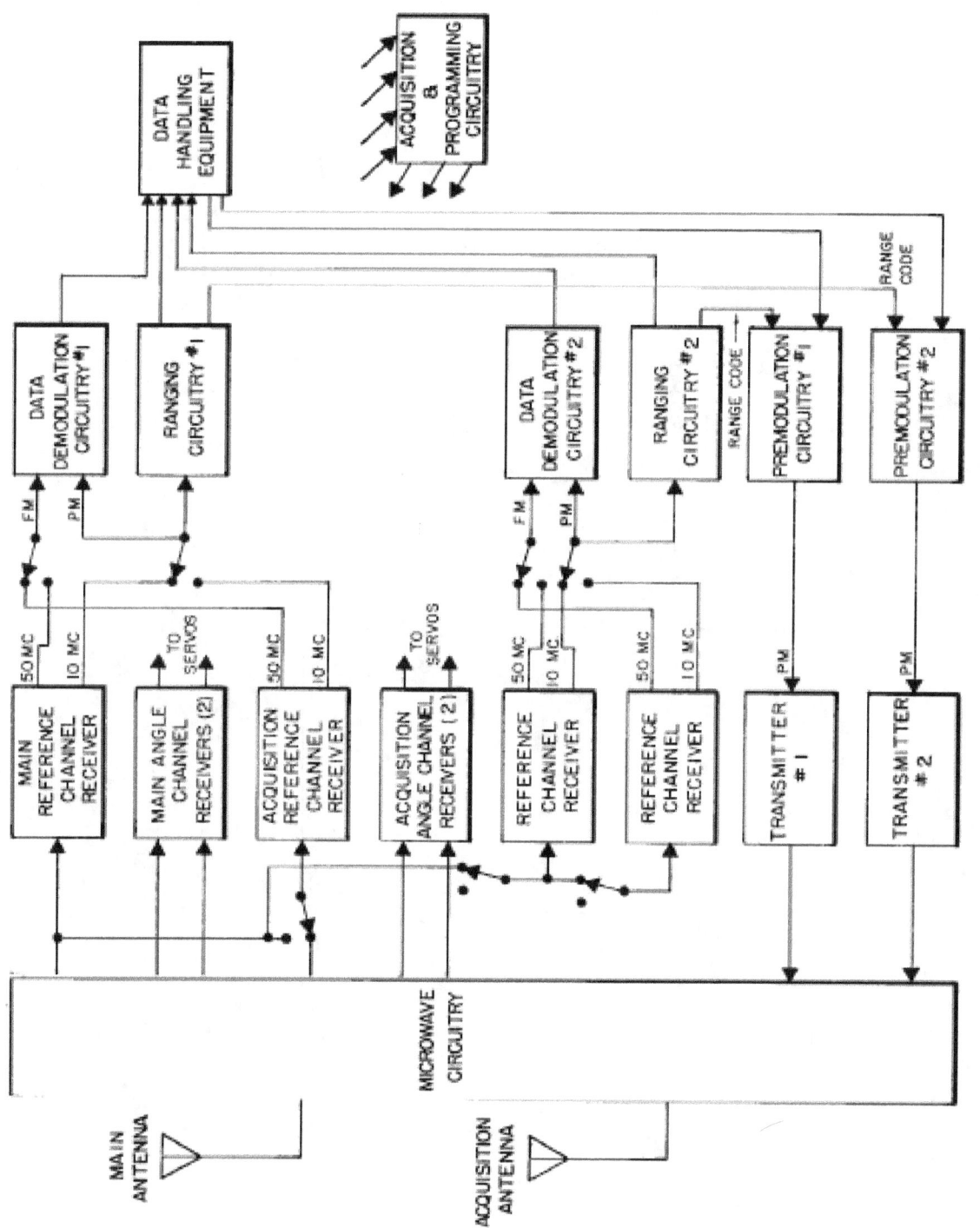

Figure 3.- The Apollo dual unified S-band ground station

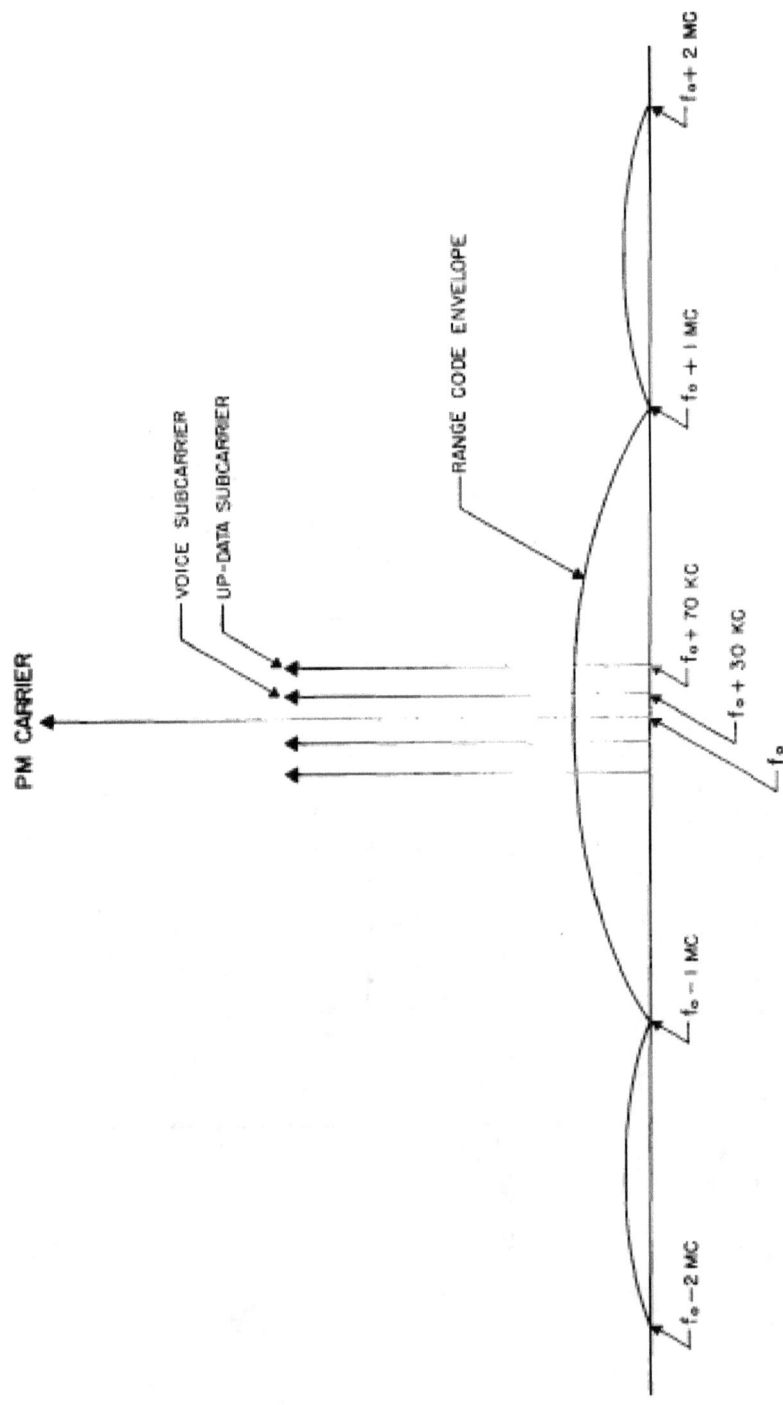

Figure 4.- The individual up-link

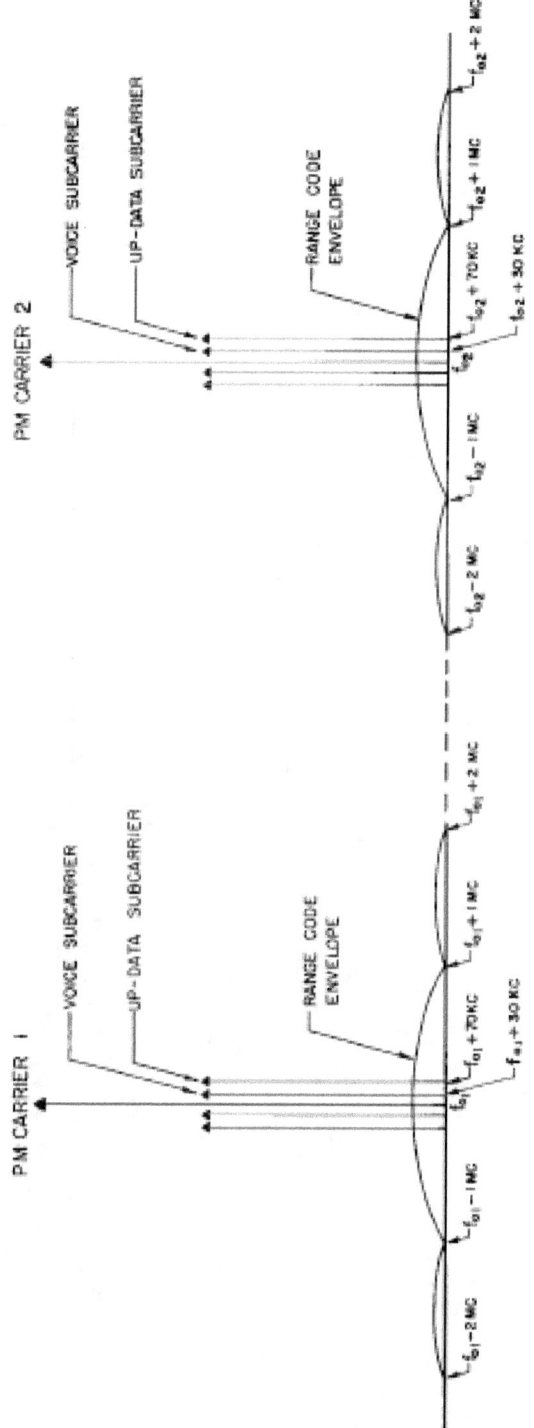

Figure 5.- The dual up-link

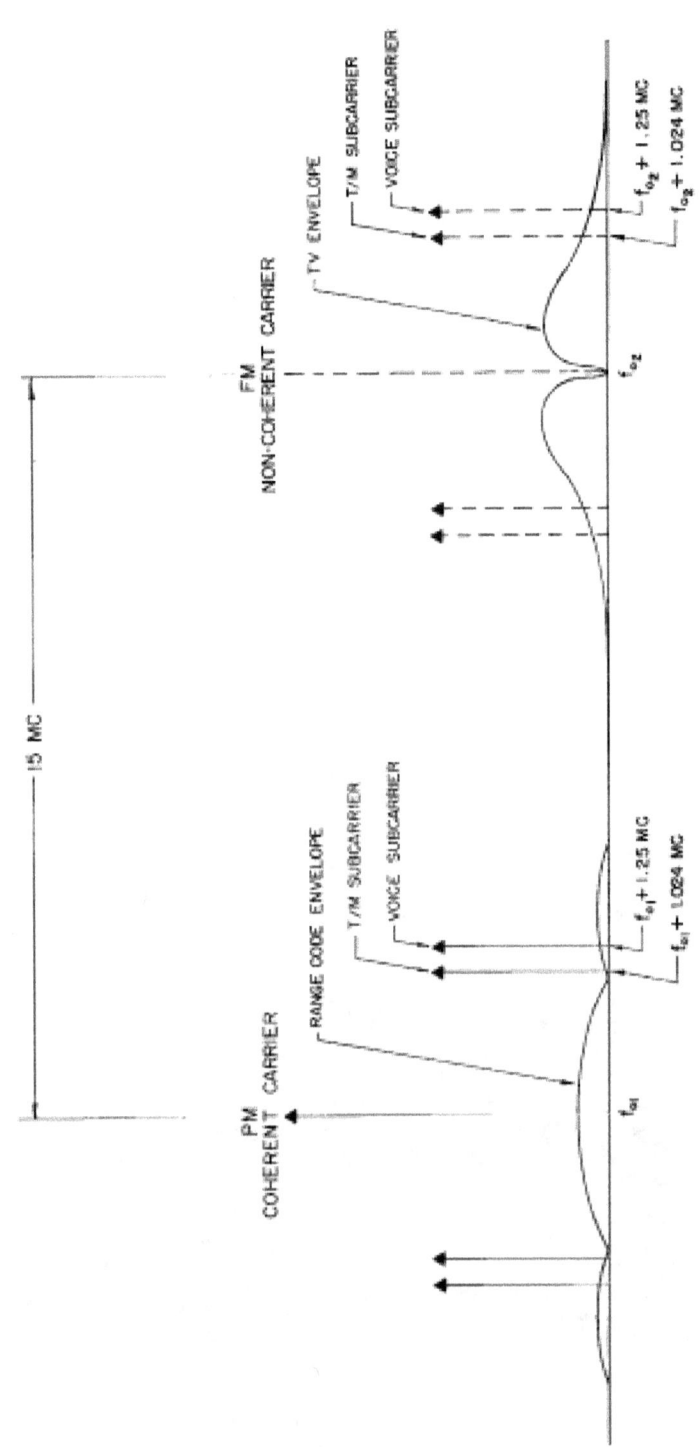

(a) CSM down-link

Figure 6.- The down-link

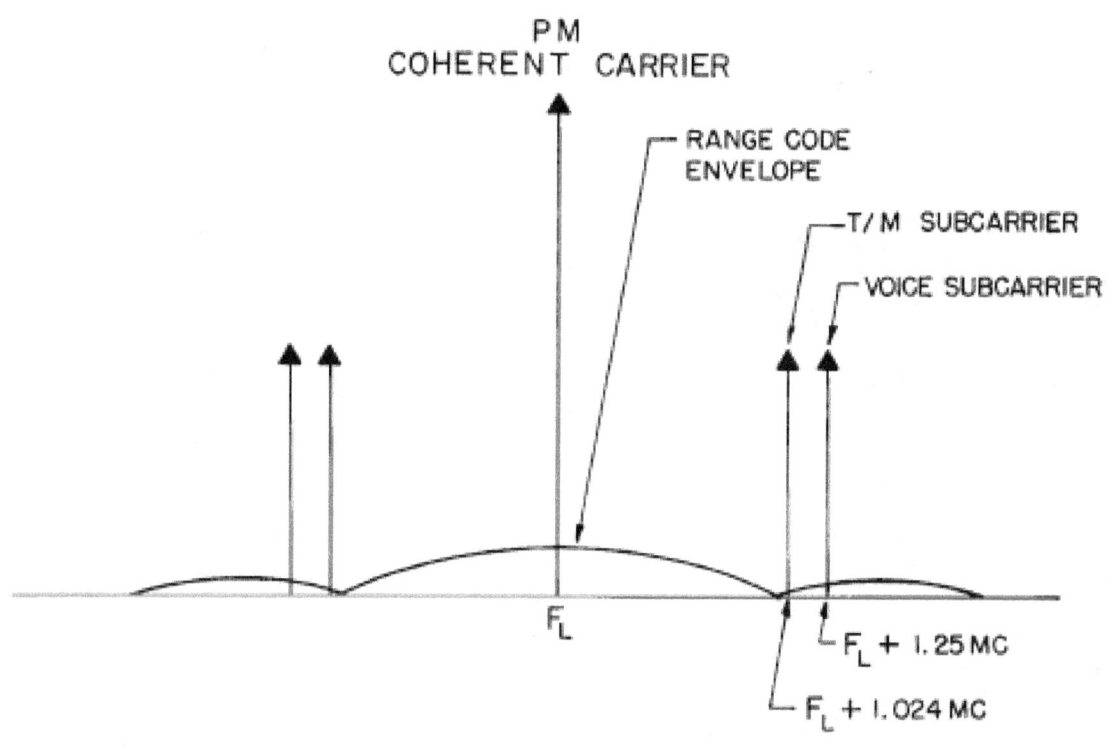

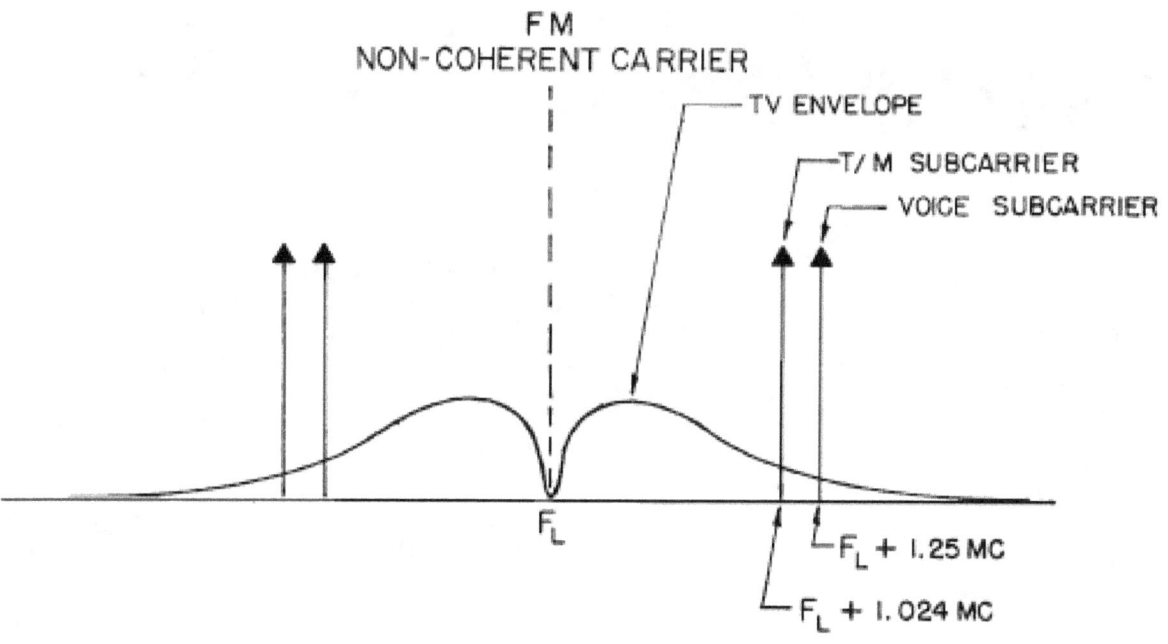

(b) LEM down-link

Figure 6.- Concluded

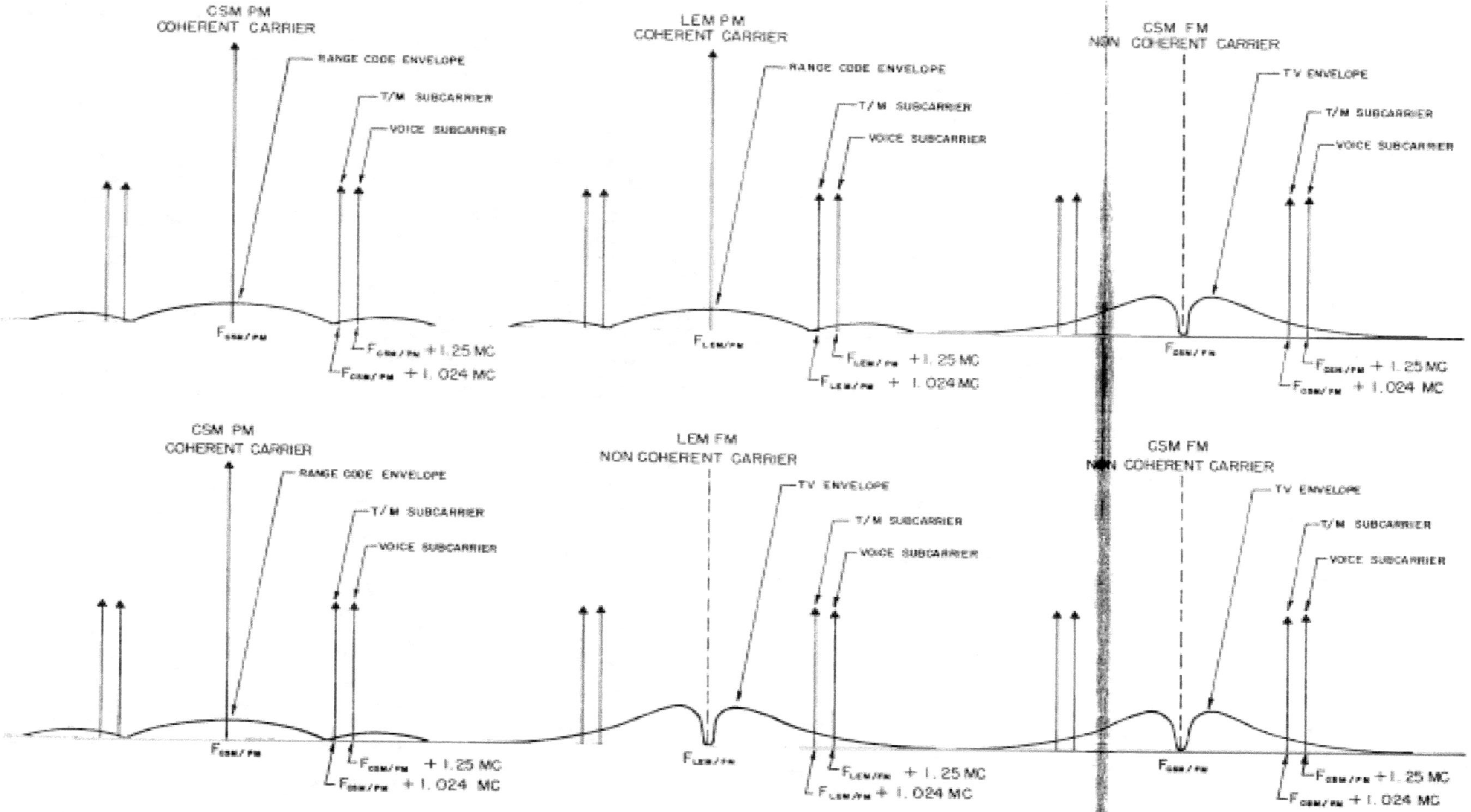

Figure 7.- The down-link composite spectrum for LEM and CSM

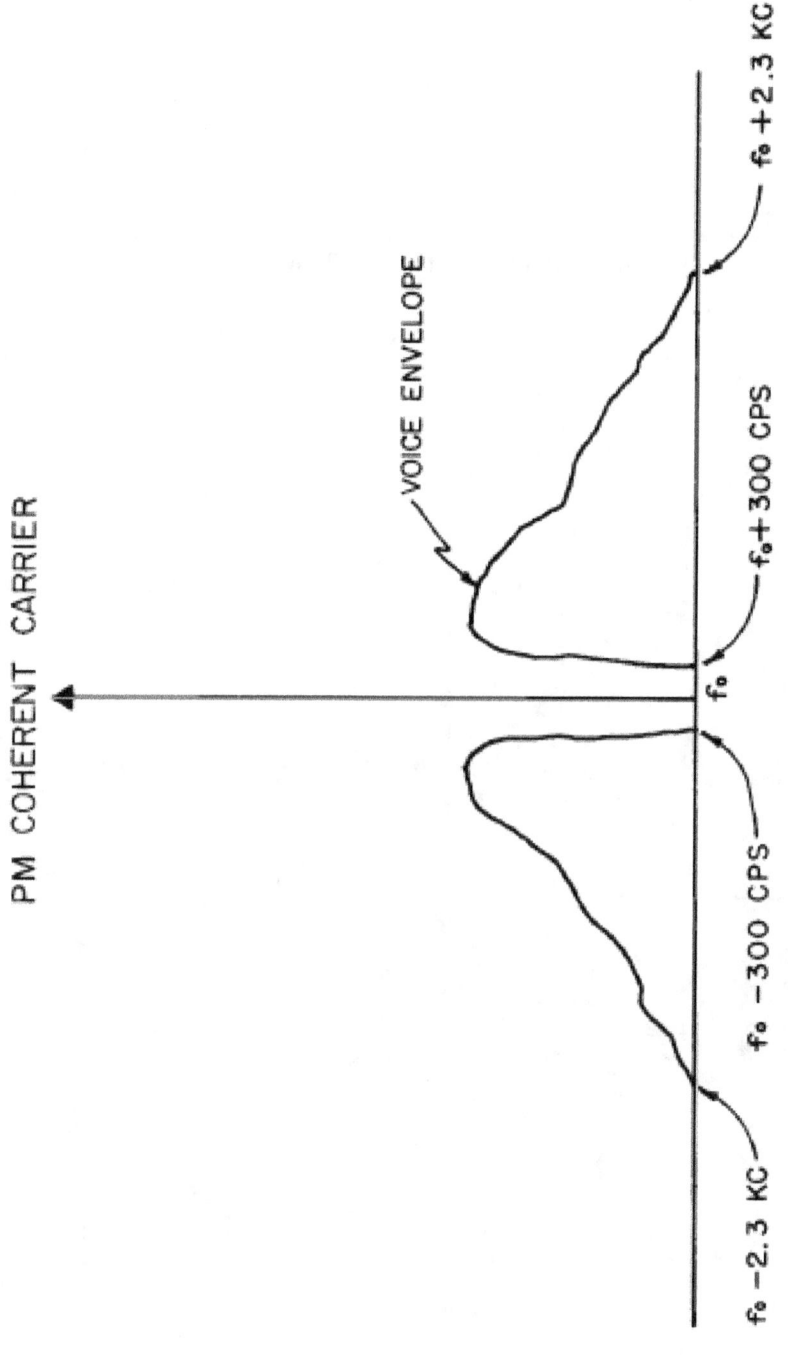

Figure 8. - The emergency voice

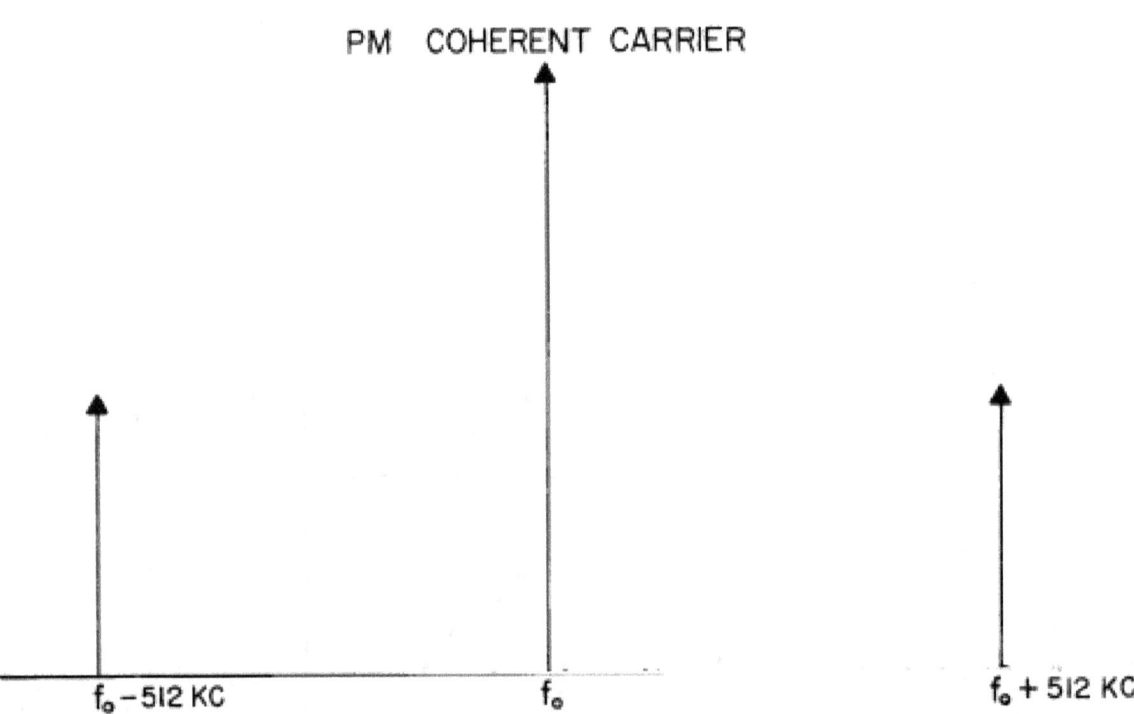

Figure 9.- The emergency key

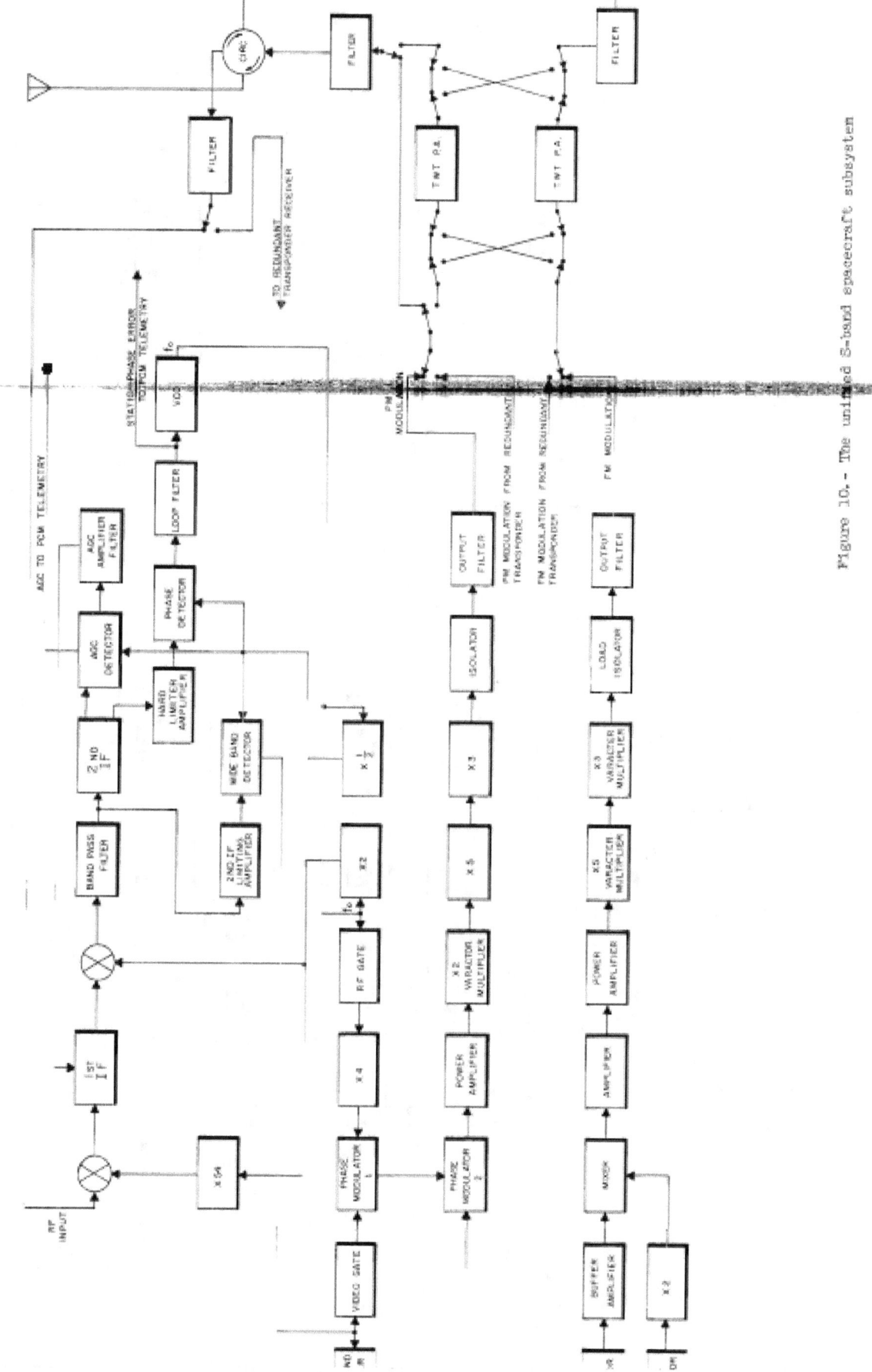

Figure 10.- The unified S-band spacecraft subsystem

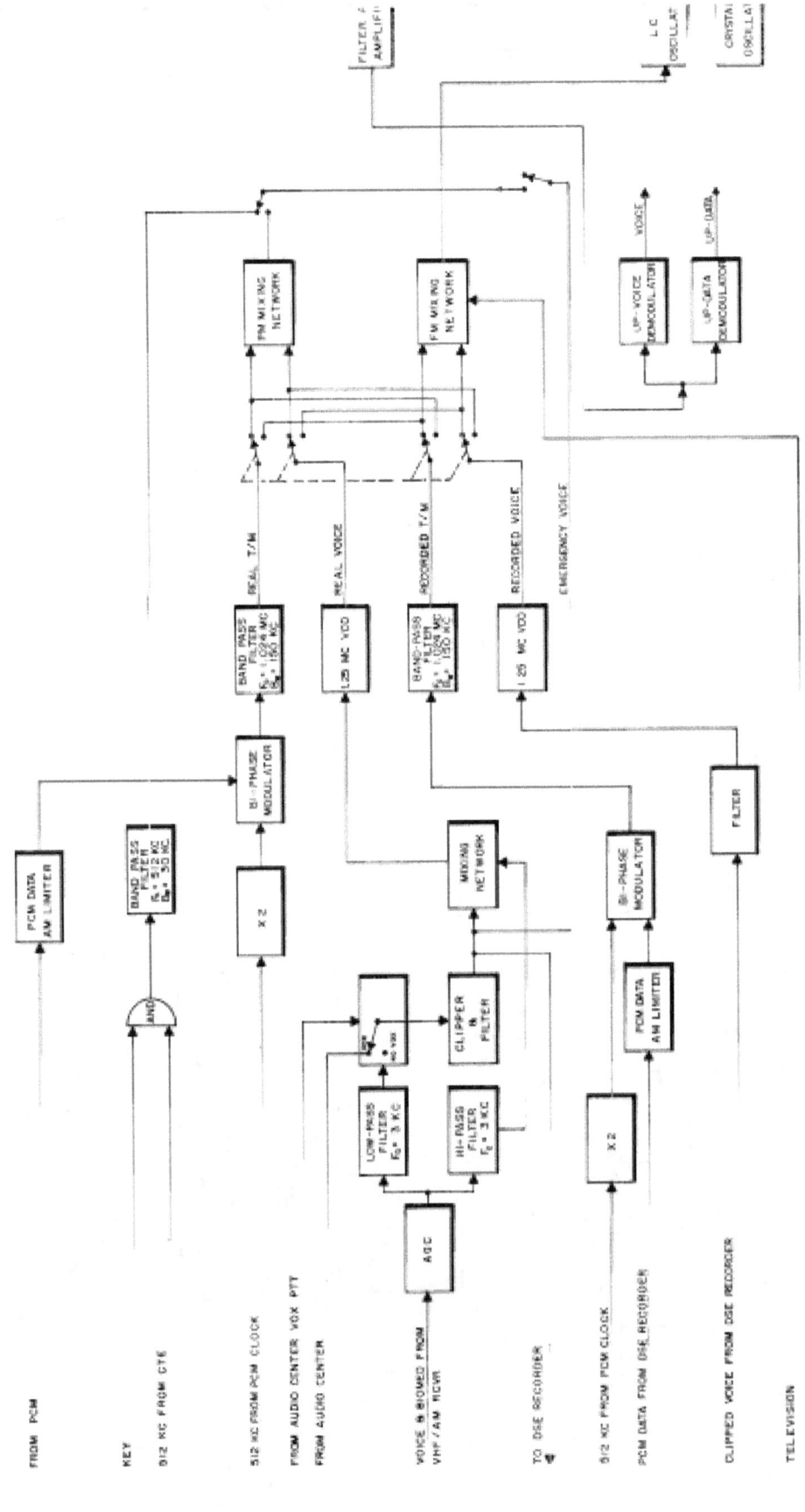

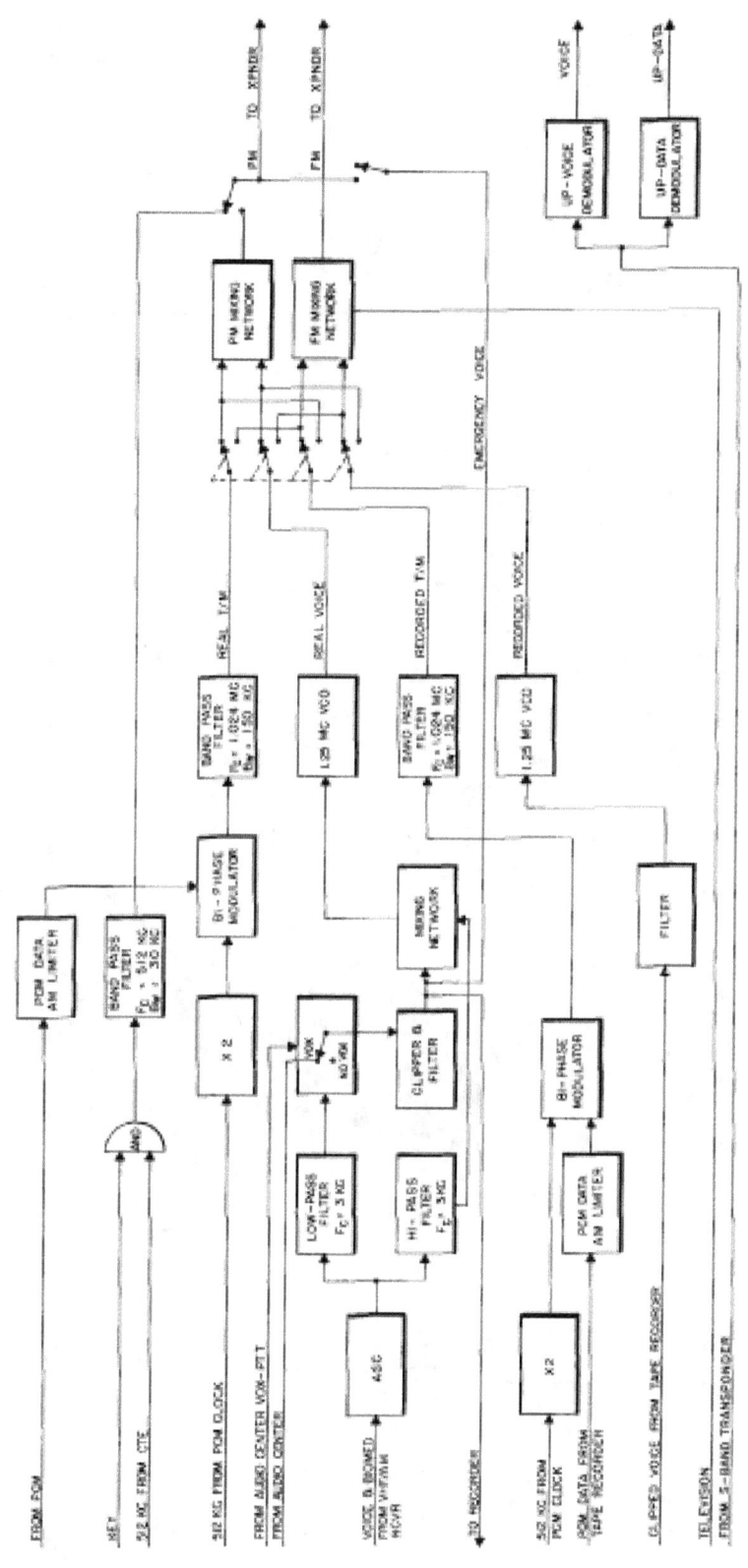

Figure 11.- The premodulation processor

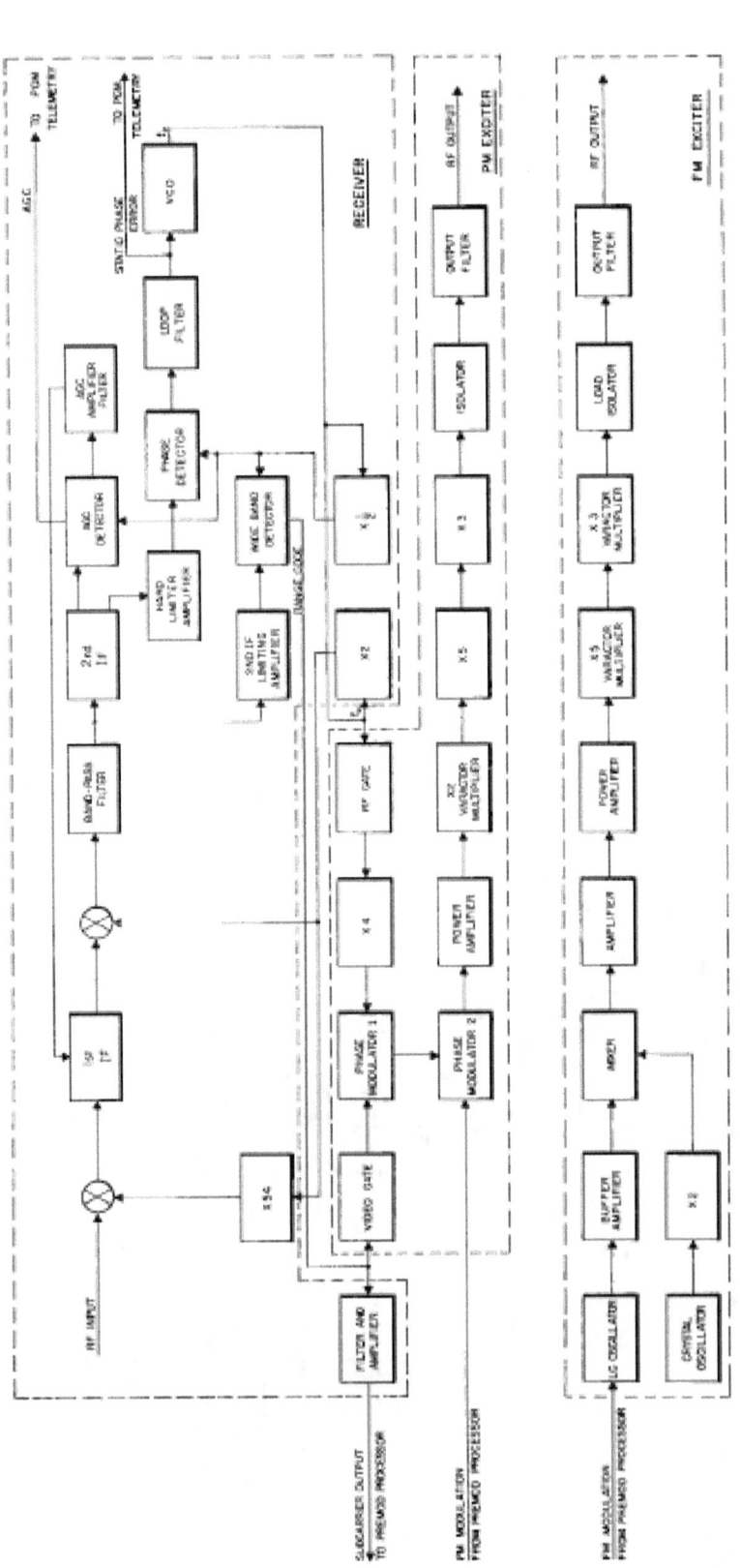

Figure 12.- The unified S-band transponder

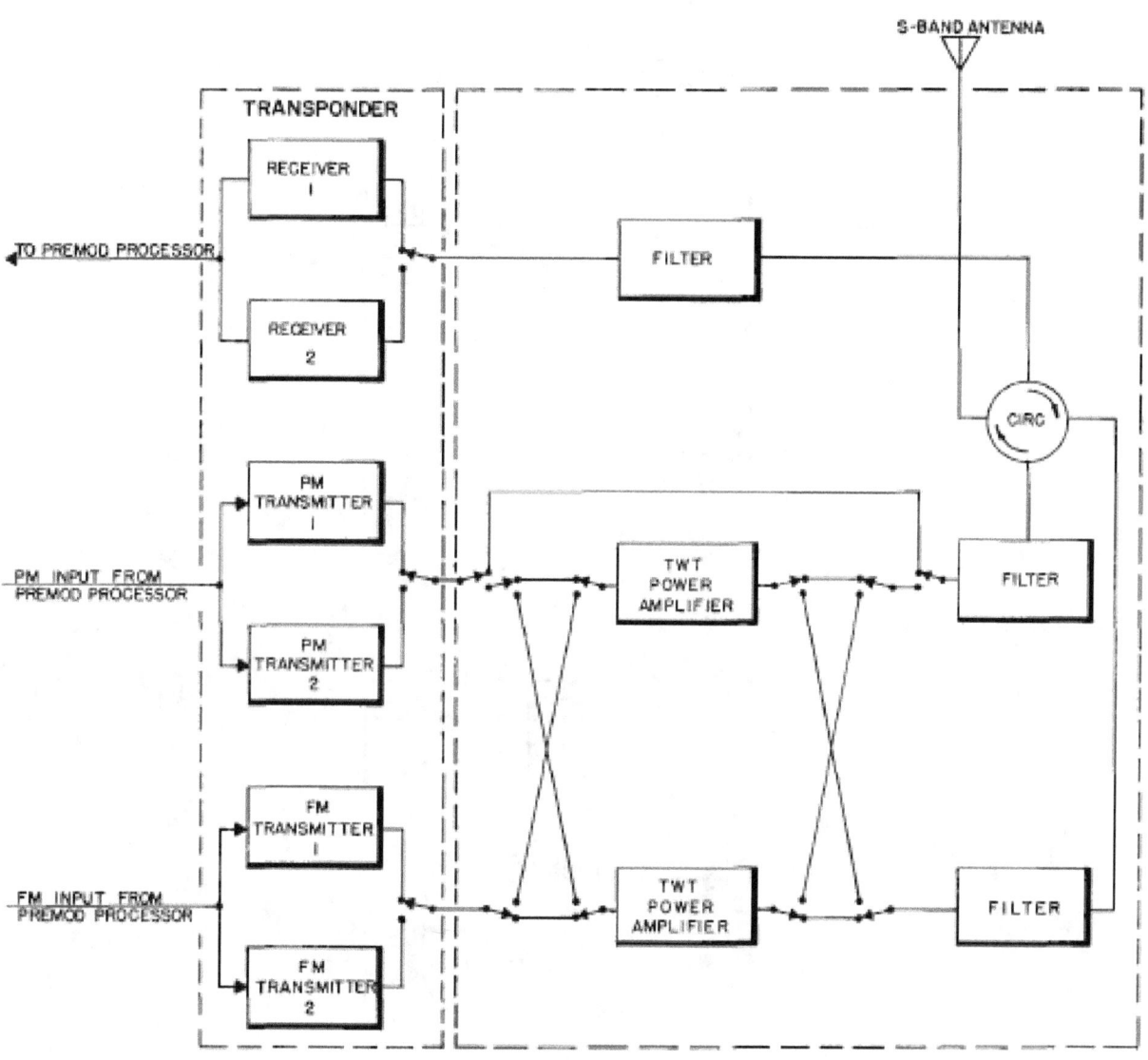

Figure 13.- The power amplifier

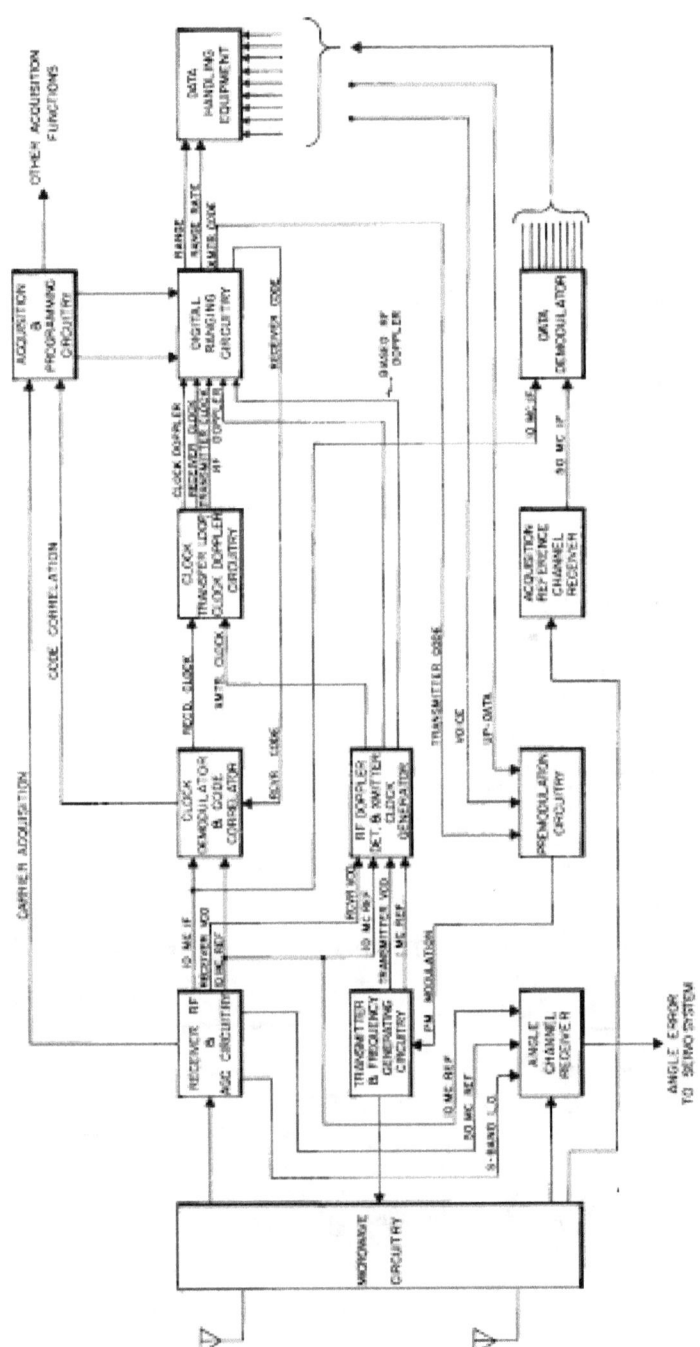

Figure 14.- The basic ground system

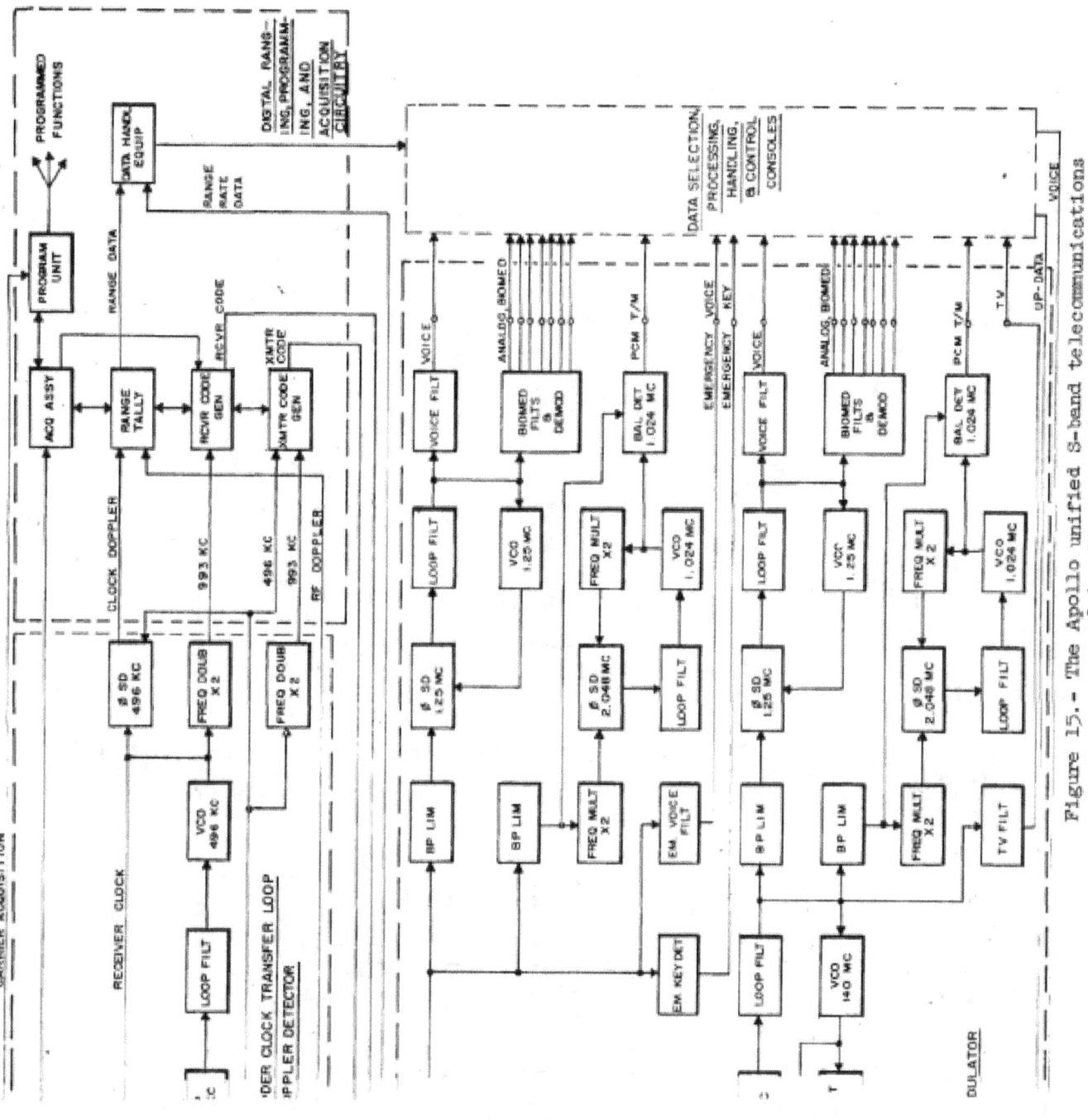

Figure 15.- The Apollo unified S-band telecommunications and tracking ground system

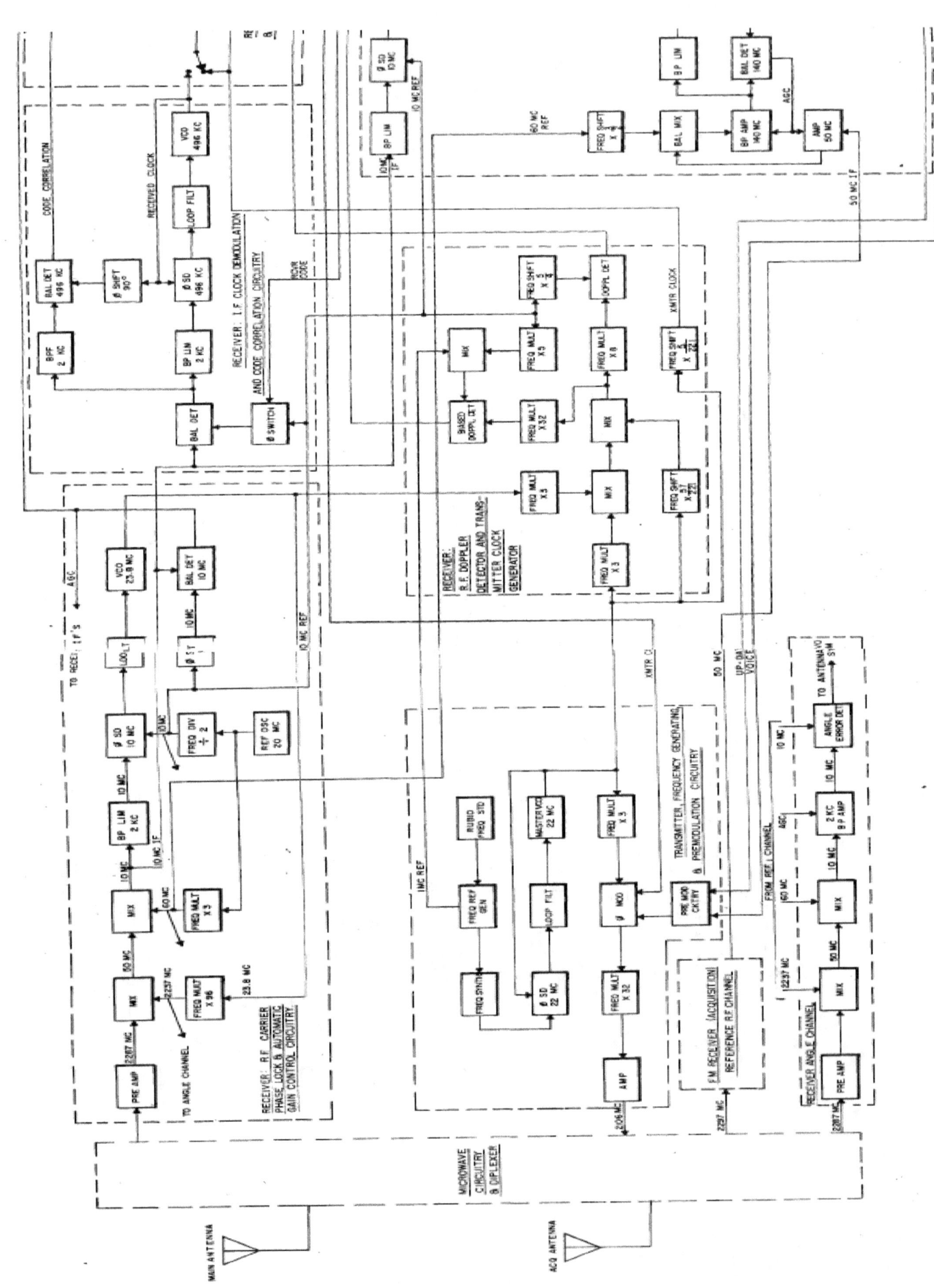

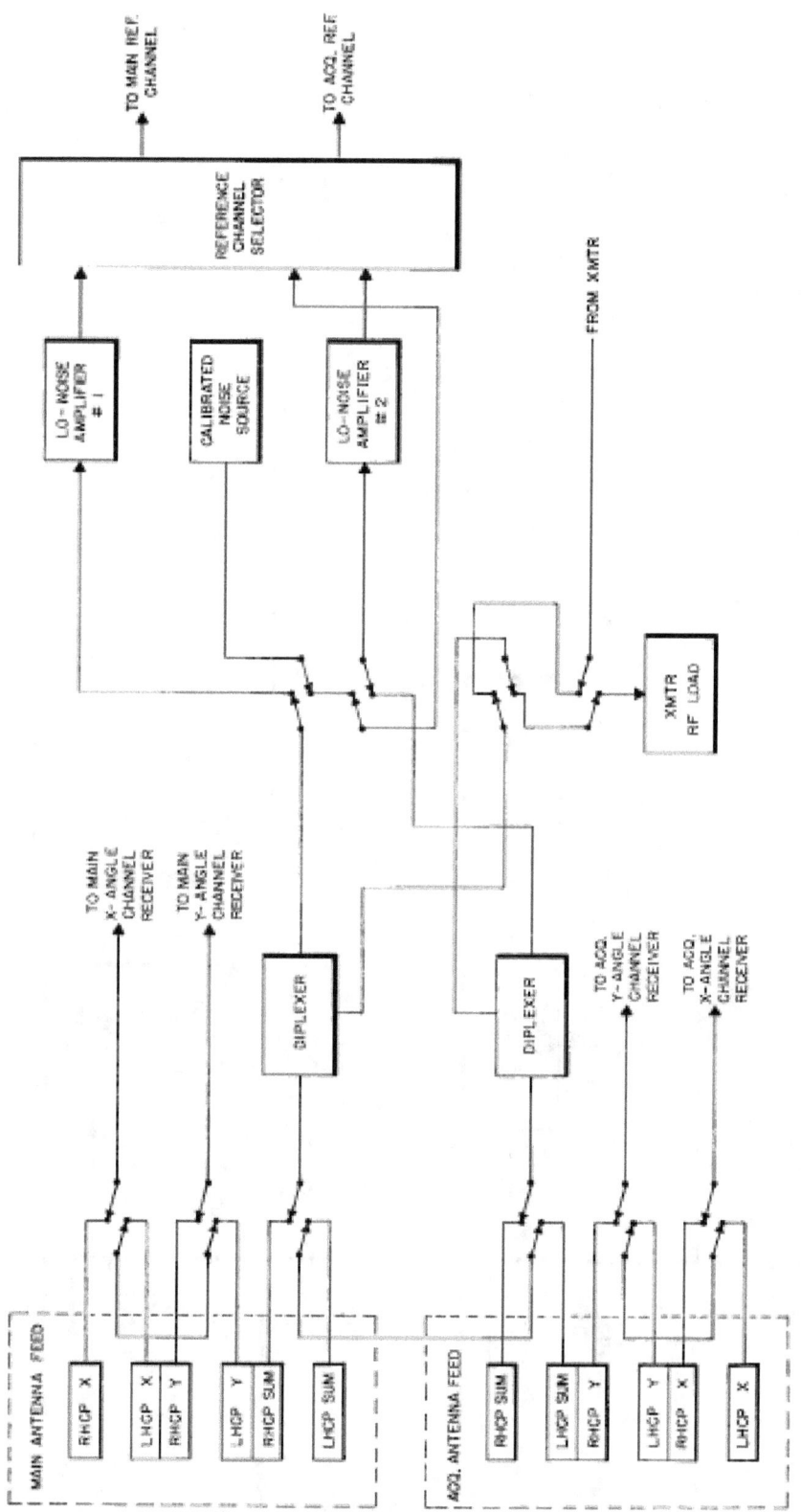

Figure 16.- The deep space station microwave circuitry

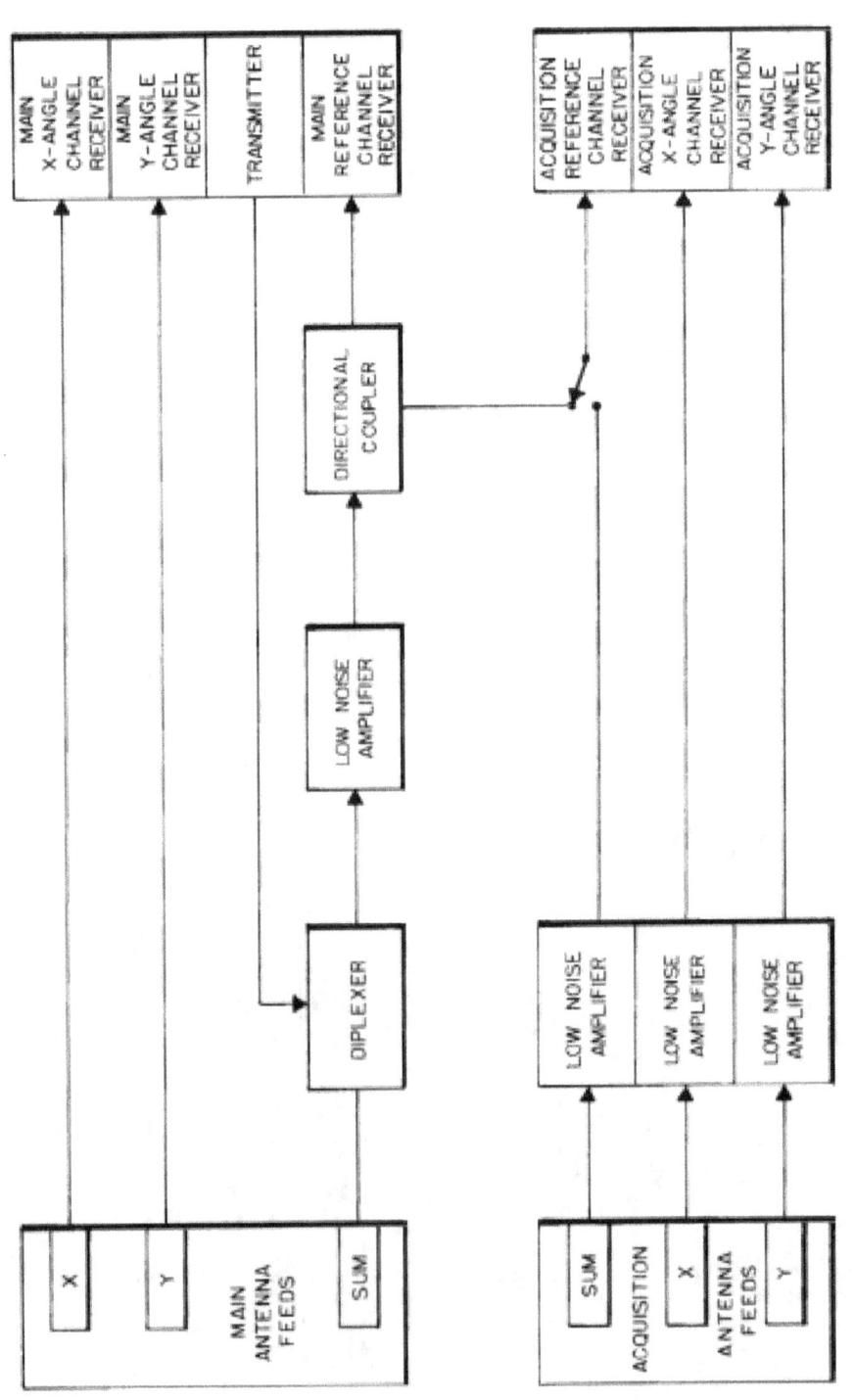

Figure 17.- The near-earth S-band ground station microwave circuitry

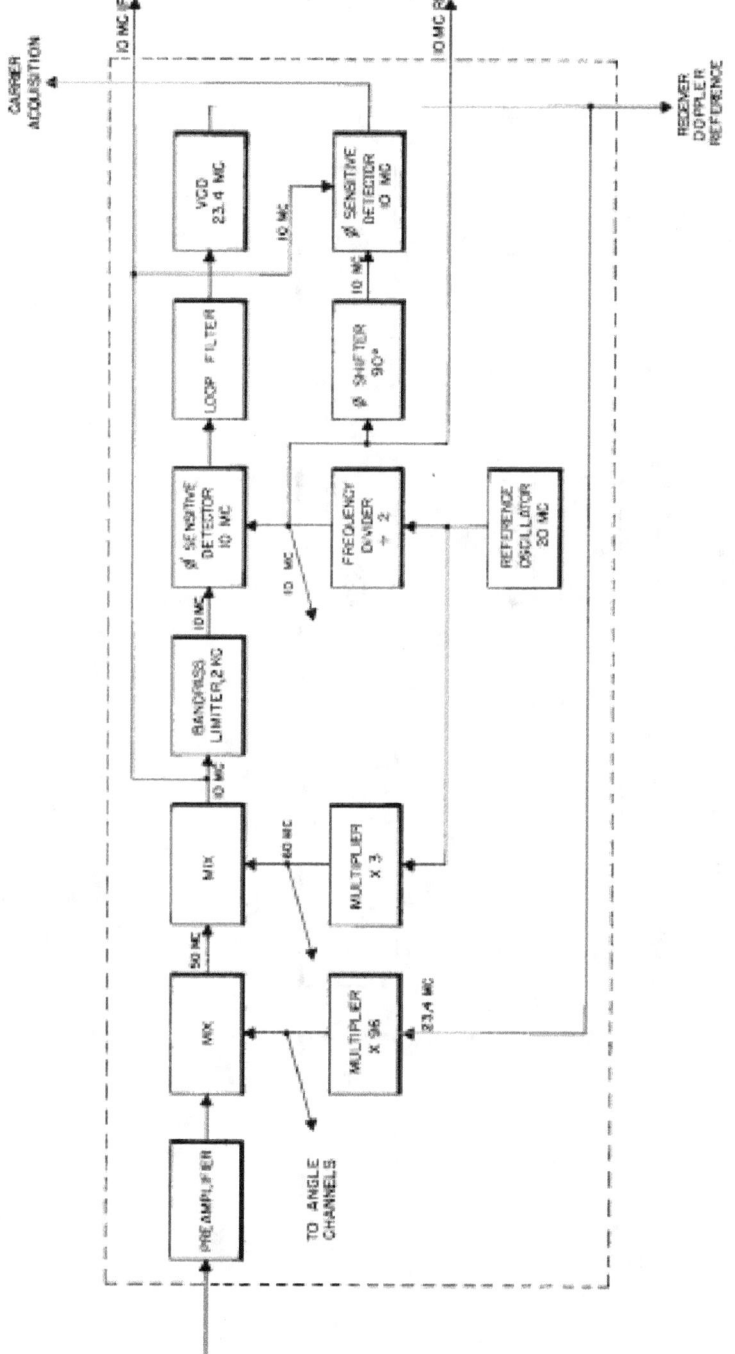

Figure 18.- The receiver RF carrier phase lock and automatic gain control circuitry

71

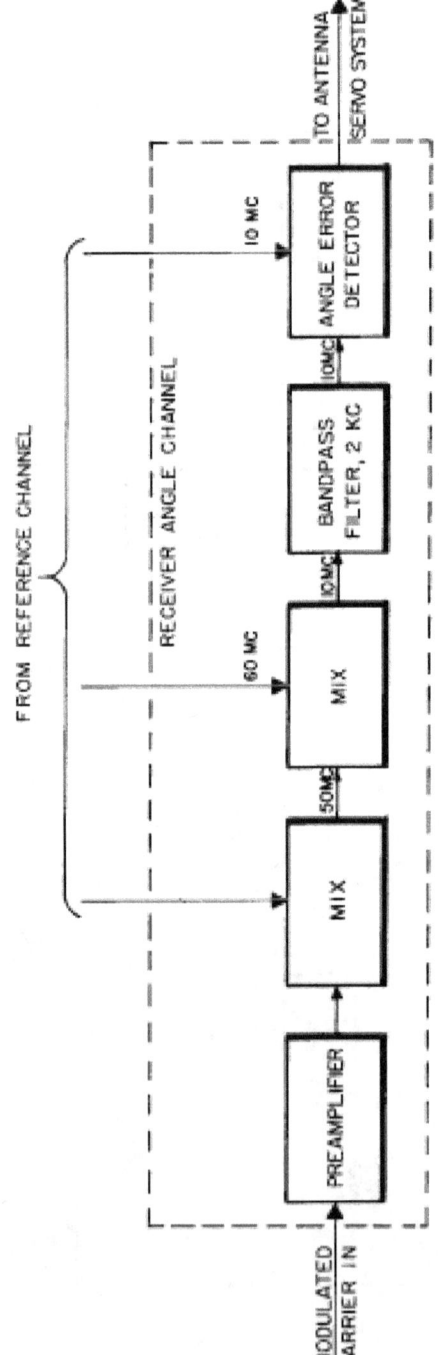

Figure 19.- The receiver angle channel

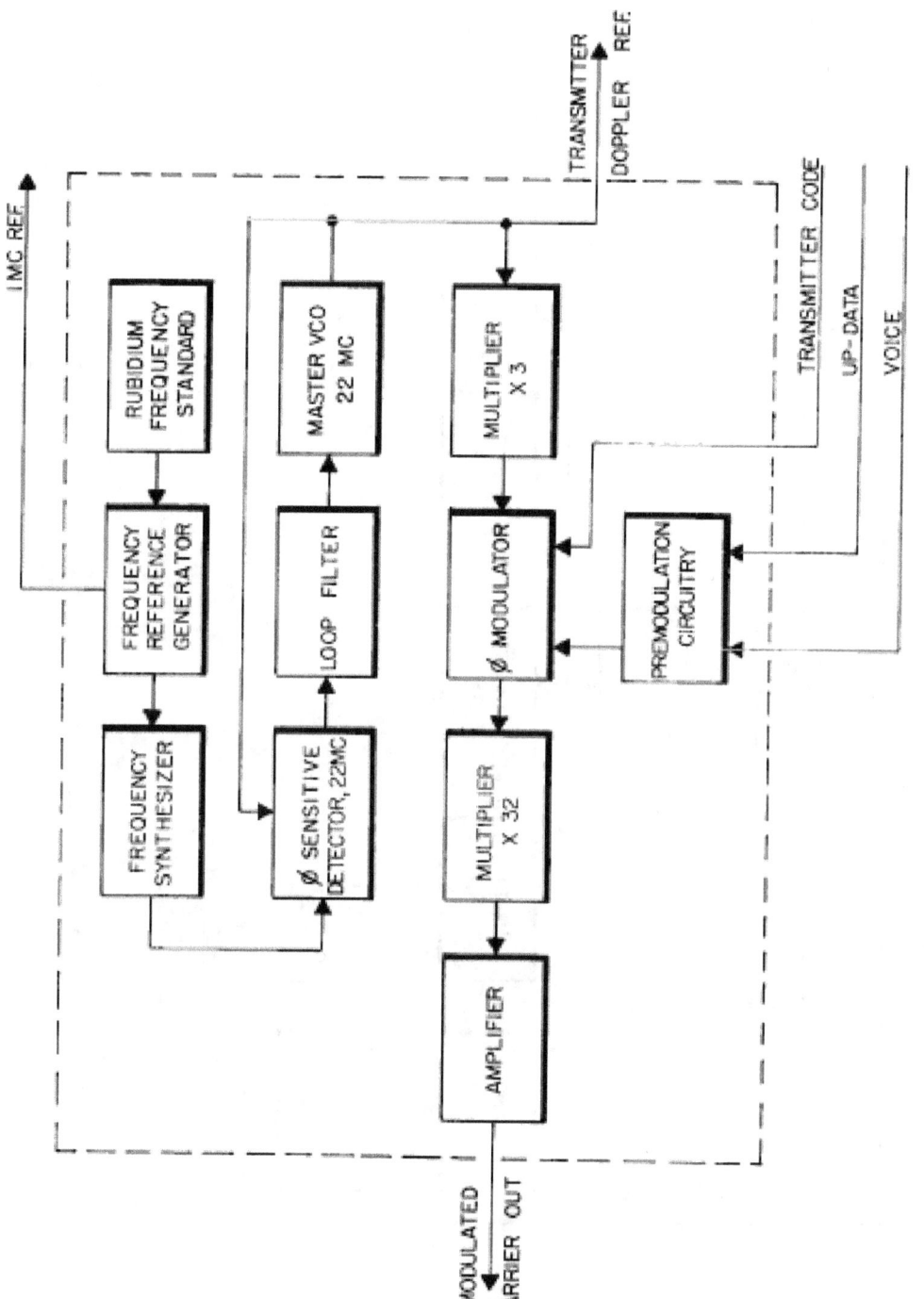

Figure 20.- The transmitter, frequency generating and pre-modulation circuitry

73

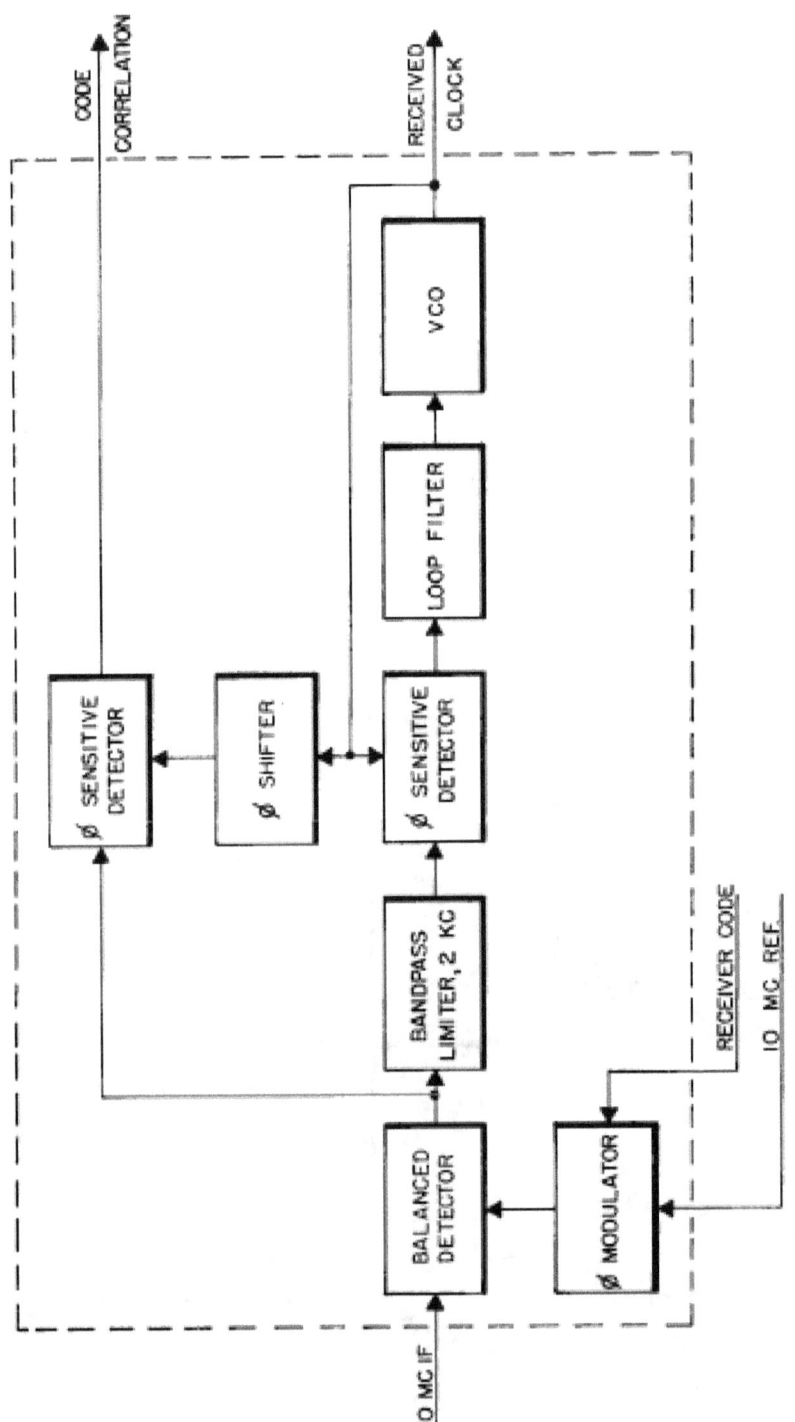

Figure 21.- The receiver IF clock demodulation and code correlation circuitry

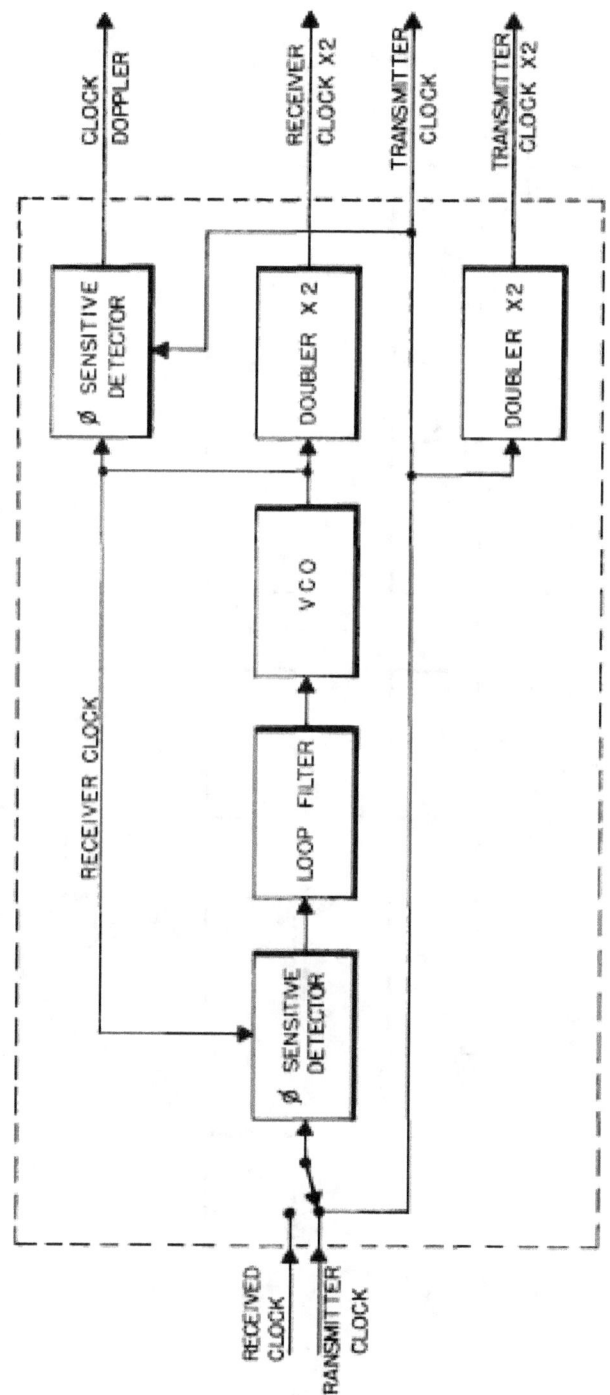

Figure 22.- The receiver coder clock transfer loop and clock Doppler detector

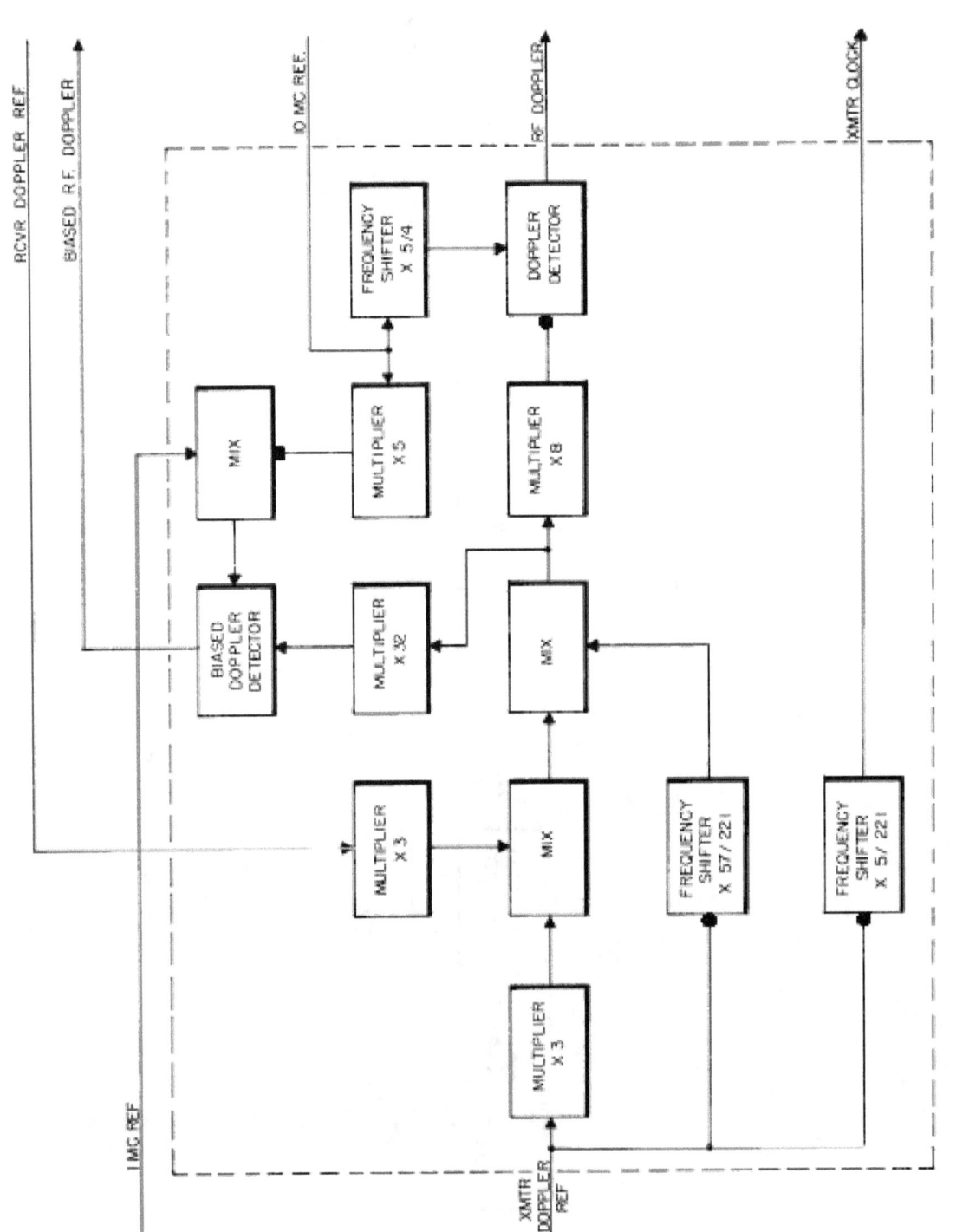

Figure 23.- The receiver RF Doppler detector and transmitter clock generator

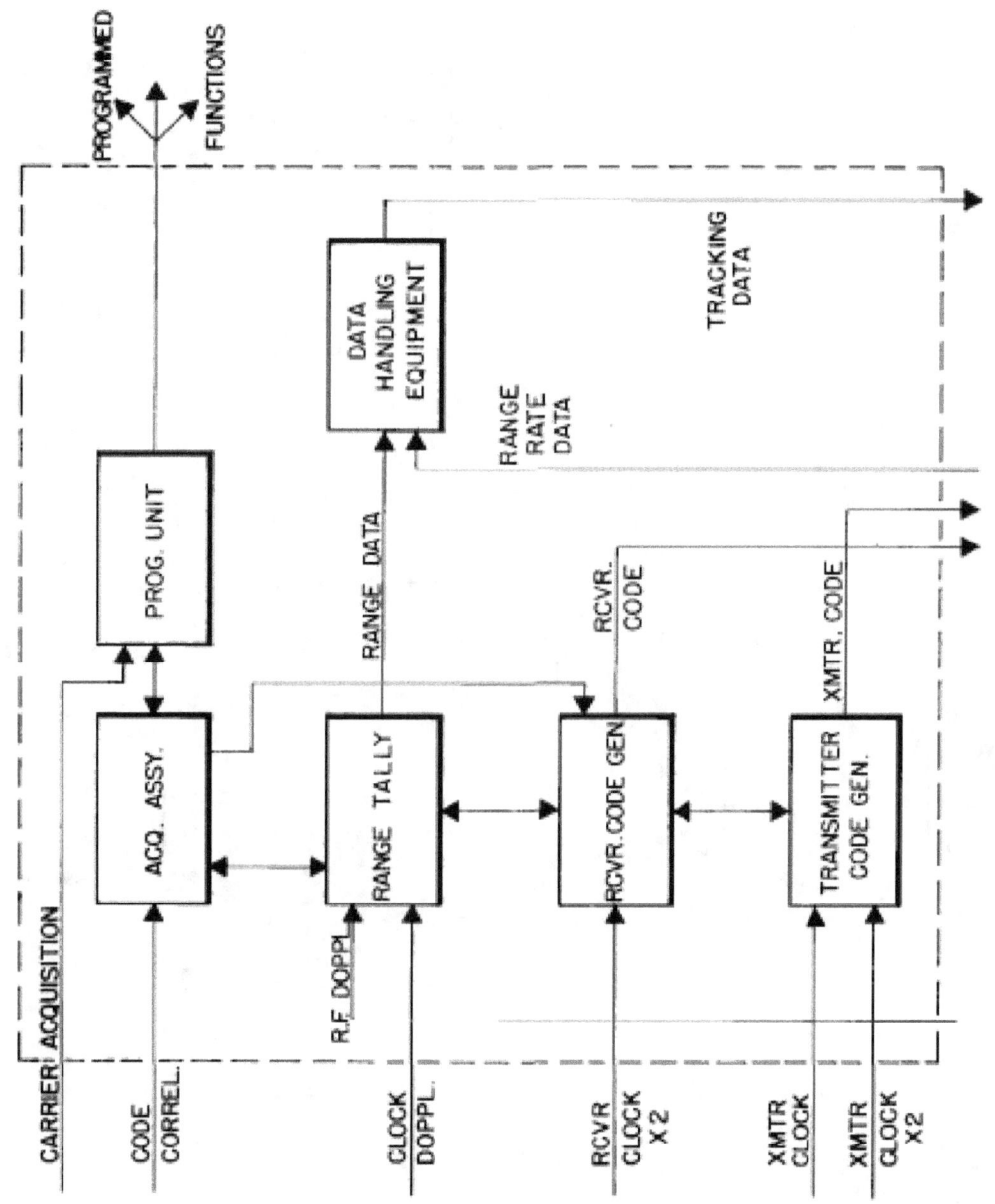

Figure 24.- The digital ranging, programing, and acquisition circuitry

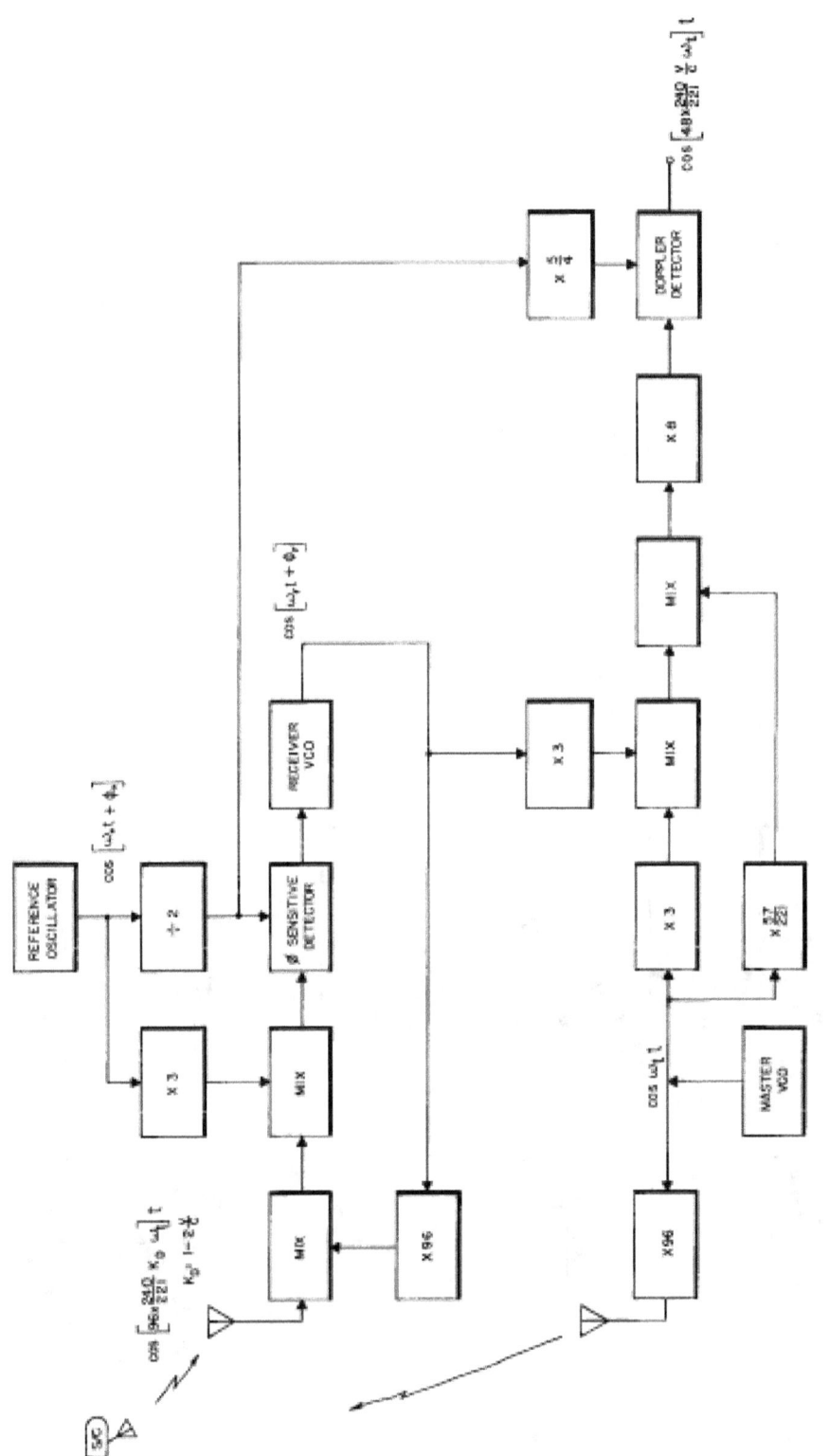

Figure 25.- The Doppler detection mechanization

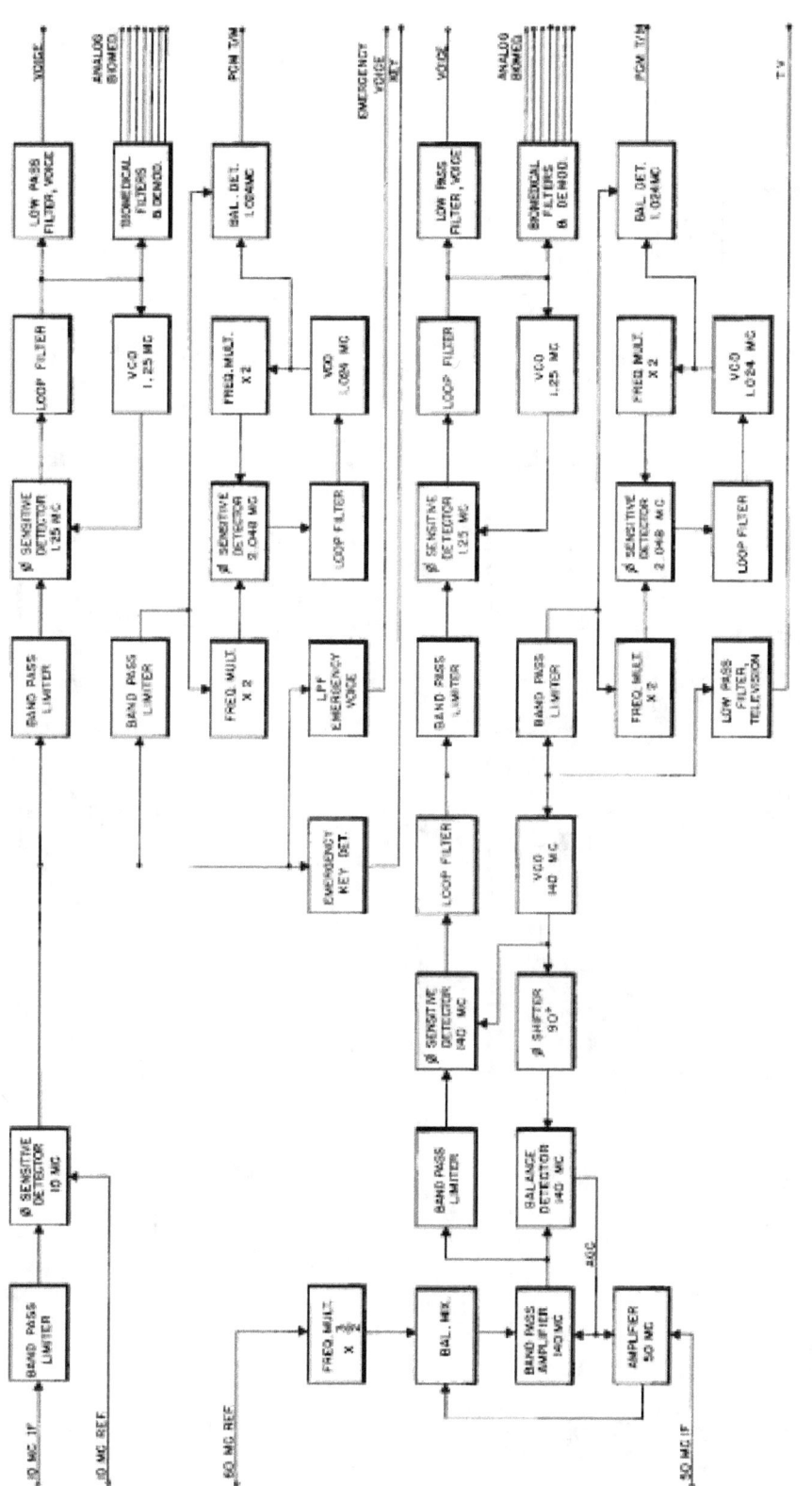

Figure 26.- The receiver data demodulator

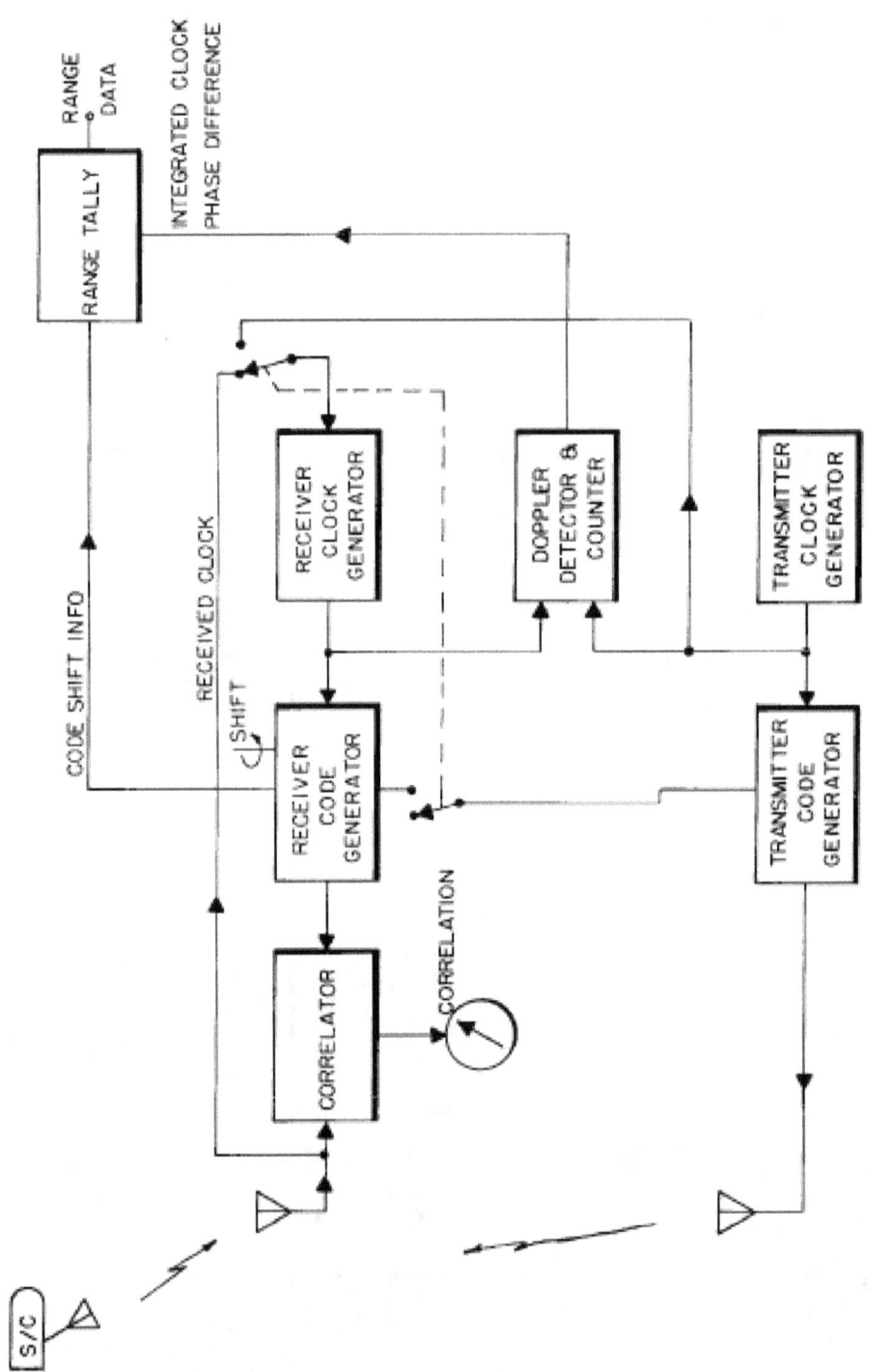

Figure 27.- The basic ranging mechanization

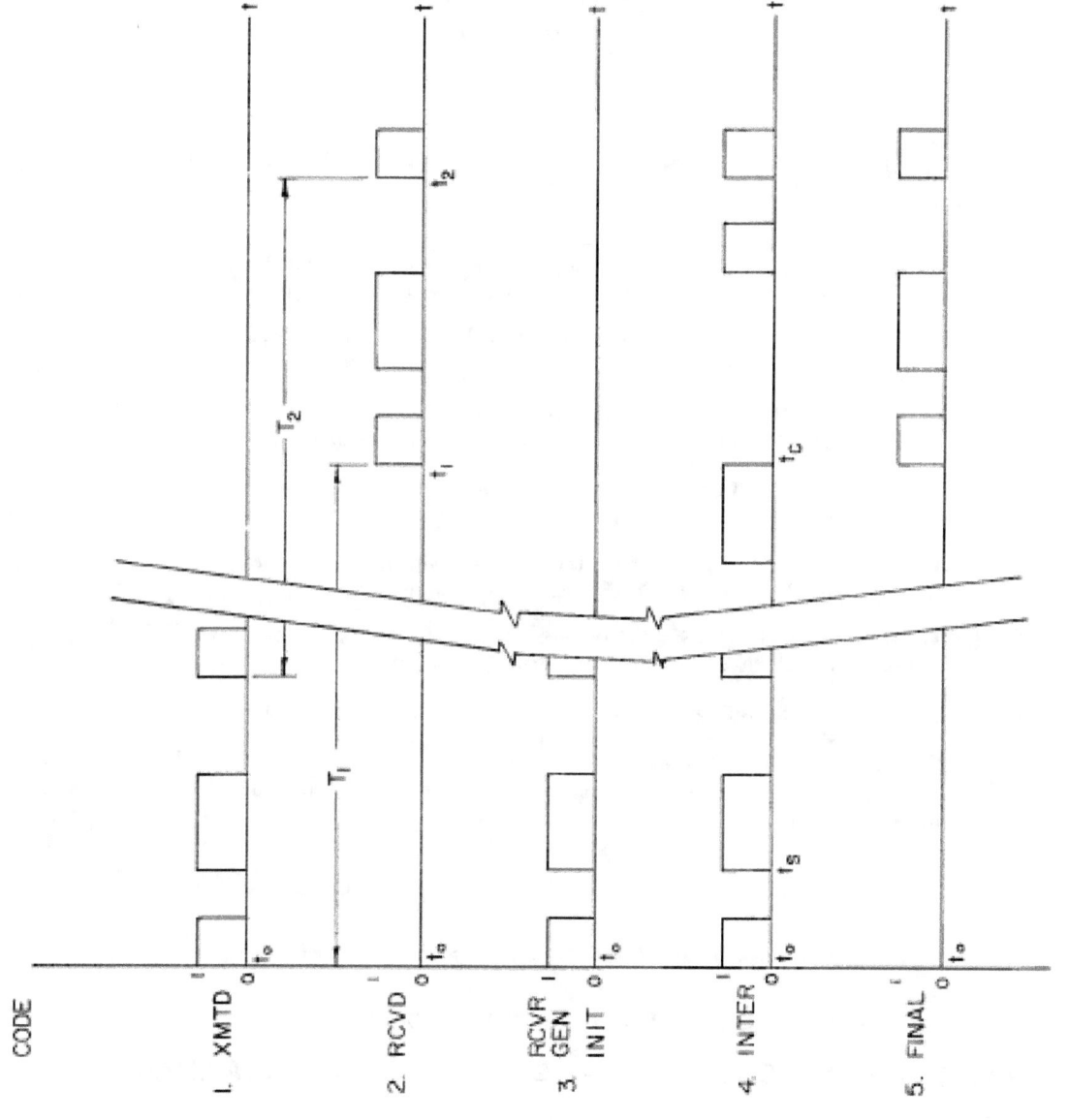

Figure 28.- The range code acquisition

"The aeronautical and space activities of the United States shall be conducted so as to contribute . . . to the expansion of human knowledge of phenomena in the atmosphere and space. The Administration shall provide for the widest practicable and appropriate dissemination of information concerning its activities and the results thereof."

—NATIONAL AERONAUTICS AND SPACE ACT OF 1958

NASA SCIENTIFIC AND TECHNICAL PUBLICATIONS

TECHNICAL REPORTS: Scientific and technical information considered important, complete, and a lasting contribution to existing knowledge.

TECHNICAL NOTES: Information less broad in scope but nevertheless of importance as a contribution to existing knowledge.

TECHNICAL MEMORANDUMS: Information receiving limited distribution because of preliminary data, security classification, or other reasons.

CONTRACTOR REPORTS: Technical information generated in connection with a NASA contract or grant and released under NASA auspices.

TECHNICAL TRANSLATIONS: Information published in a foreign language considered to merit NASA distribution in English.

TECHNICAL REPRINTS: Information derived from NASA activities and initially published in the form of journal articles.

SPECIAL PUBLICATIONS: Information derived from or of value to NASA activities but not necessarily reporting the results of individual NASA-programmed scientific efforts. Publications include conference proceedings, monographs, data compilations, handbooks, sourcebooks, and special bibliographies.

Details on the availability of these publications may be obtained from:

SCIENTIFIC AND TECHNICAL INFORMATION DIVISION
NATIONAL AERONAUTICS AND SPACE ADMINISTRATION
Washington, D.C. 20546

www.ingramcontent.com/pod-product-compliance
Lightning Source LLC
Chambersburg PA
CBHW081734170526
45167CB00009B/3812